T5-AGM-141

THALASSAEMIA

THALASSAEMIA

The Biography

David Weatherall

OXFORD
UNIVERSITY PRESS

Great Clarendon Street, Oxford OX2 6DP
Oxford University Press is a department of the University of Oxford.
It furthers the University's objective of excellence in research, scholarship,
and education by publishing worldwide in

Oxford New York

Auckland Cape Town Dar es Salaam Hong Kong Karachi
Kuala Lumpur Madrid Melbourne Mexico City Nairobi
New Delhi Shanghai Taipei Toronto

With offices in
Argentina Austria Brazil Chile Czech Republic France Greece
Guatemala Hungary Italy Japan Poland Portugal Singapore
South Korea Switzerland Thailand Turkey Ukraine Vietnam

Published in the United States
by Oxford University Press Inc., New York

First published 2010

British Library Cataloguing in Publication Data
Data available

Library of Congress Cataloging in Publication Data
Library of Congress Control Number: 2010930323

Typeset by SPI Publisher Services, Pondicherry, India
Printed in Great Britain
on acid-free paper by
Clays Ltd., St Ives Plc

ISBN 978-0-19-956560-3

1 3 5 7 9 10 8 6 4 2

PREFACE

The thalassaemias are among the most common genetic diseases and affect thousands of children throughout the world. This short account of the history of this disorder and how the elucidation of its underlying causes helped to lead the way towards the era of molecular medicine is based in part on a more extensive account in an earlier book, *The Thalassaemia Syndromes*, and several other essays and reviews on the same theme. For readers who wish to explore aspects of this field in more detail, or its extensive literature, there is a short bibliography to point them in the right direction. For those who are unfamiliar with current genetics, there is also a short glossary.

In tracing this complex story I have had the help of many colleagues over the years, notably Maxwell Wintrobe, Ida Bianco, Robin Bannerman, Vernon Ingram, and John Clegg. I would also like to acknowledge the help of Jeanne Packer and Liz Rose in preparing this manuscript, Lady Brabourne for giving permission to reproduce the picture of Lady Mountbatten and Jaspir Thapa, Wolf Zeulzer for the photograph of Thomas Cooley, Elizabeth Letsky for the photograph of the thalassaemic child, and Blackwell/Wiley for permission to reproduce other figures from *The Thalassaemia Syndromes*.

CONTENTS

LIST OF ILLUSTRATIONS

LIST OF ILLUSTRATIONS

PROLOGUE

Today's visitors to Singapore who wish to escape from the hustle and bustle of its extraordinary city and explore its quieter suburbs may chance on the Alexandra Hospital, a collection of tall and elegant buildings with deep verandas, set among tall trees and luscious greenery. Built by the British in 1938, it has a fascinating history. In February 1942 it was the site of a brutal massacre, when over 200 patients and staff were killed by invading Japanese forces. Between 1948 and 1960, during the Emergency—the war in Malaya between the Chinese communists and Commonwealth forces—as the British Military Hospital it acted as the main centre for casualties who could not be managed in the smaller hospitals in Malaya and for the British and Commonwealth forces and their families. In 1971 it was finally handed over to the Singapore government and is now an extremely fine teaching hospital.

As a recently qualified doctor, I arrived in Singapore late in 1958 to serve my two years of national service. This was equally divided between a year at the Alexandra Hospital in Singapore and a year at the military hospital in Taiping, North Malaya, where the remaining Chinese communist terrorists were being hunted down in the jungle on the border with Thailand. I spent most of my time at the Alexandra Hospital in Singapore as medical officer in charge of the children's ward, and it was there that I first met a remarkable young Nepalese girl called Jaspir Thapa.

Jaspir's father was a sergeant in a Ghurkha regiment and he had brought his family down from Nepal when he was posted to Malaya during the Emergency. At the age of 3 months Jaspir was investigated at the British Military Hospital in Kinrara and found to be severely anaemic. No cause could be found, and she was kept alive by regular blood transfusion over subsequent years. When her father was posted to Singapore, she became a patient at the Alexandra Hospital, and it was there that I struggled to try to maintain her on transfusions. Despite extensive studies, I had no idea why Jaspir was chronically anaemic until a chance meeting with Frank Vella, a Maltese biochemist who worked at the University of Singapore. He brought my attention to a paper that had been published a few years earlier in Thailand with the bizarre title 'Mediterranean Anemia in Thailand'.* This report described patients with a condition called thalassaemia, which, until then, was thought to be mainly restricted to Mediterranean populations. In 1958 new techniques became available for a more accurate diagnosis of thalassaemia, and we were able to show that Jaspir was undoubtedly suffering from this condition, which was inherited from her parents. Under the heading 'Thalassaemia in a Ghurkha Family', we published these findings in the *British Medical Journal*. This was not, as it turned out, an auspicious start to a research career, since I was nearly court-martialled for publishing data about British military personnel without

* Readers who are unfamiliar with the haematological literature should be reassured that this book is not full of misprints. In the USA the words 'anaemia' and 'thalassaemia' are spelled 'anemia' and 'thalassemia'. According to H. L. Mencken's *The American Language*, the reason is obscure.

getting permission from the War Office! And, I was told, it is in any case extremely bad form to tell the world that soldiers in one of our most outstanding regiments carry bad genes.

Jaspir's father was due to complete his period of service with the Ghurkhas at the beginning of 1960, and the family were aware that when they returned to their isolated village in Nepal Jaspir would almost certainly die within a short time because there were no facilities for further transfusion. During the late 1950s Lady Mountbatten, as part of her long-standing work for the St John Ambulance Brigade and the Red Cross, made frequent visits to Malaya and Singapore and showed a particular interest in the children's ward at Alexandra Hospital, where she met Jaspir on several occasions and obviously became very attached to this increasingly lively young child.

The last press photograph of Lady Mountbatten before her death in February 1960 was taken while she was playing with Jaspir in the children's ward in Singapore. Lord Mountbatten became so fond of this picture that he had a local artist convert it into a drawing (Figure 1). Some years later this appeared in a British newspaper under a caption that contained a number of inaccuracies about Jaspir. I wrote to Lord Mountbatten at the time, enclosing a photograph and an account of Jaspir's life and illness. Later I had a charming reply from him (Figure 2), saying that this was undoubtedly Jaspir and that he would have the inaccuracies corrected. The picture now hangs in the Mountbatten family home, Broadlands, in Romsey, Hampshire, a moving testimony to two remarkable though very different people, both of whom appear to be so happy shortly before their untimely deaths.

This short biography of thalassaemia describes how a disease that was at first thought to be rare and confined to

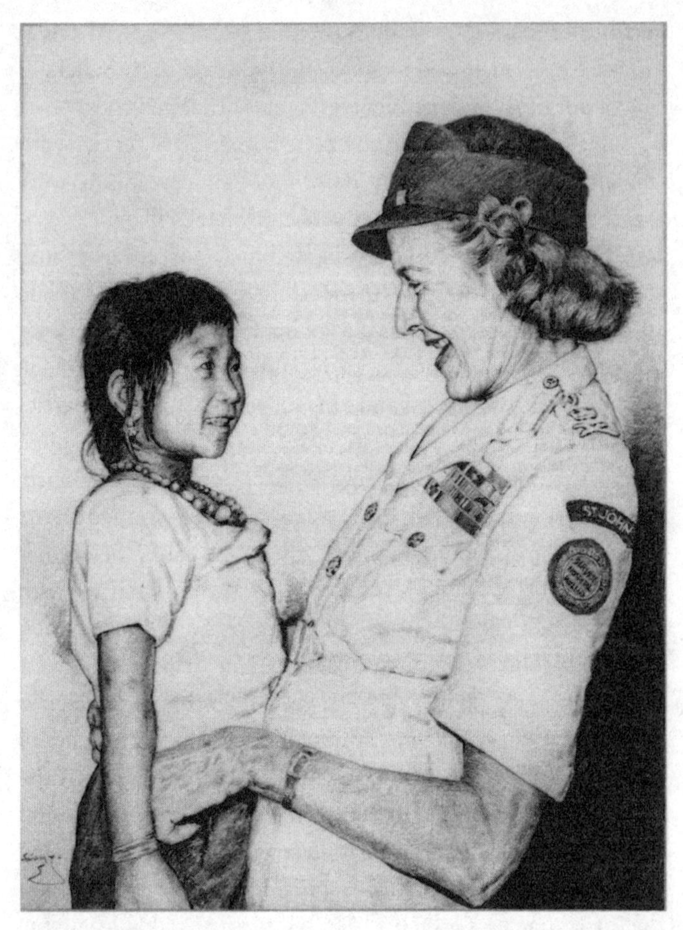

1. Lady Mountbatten with Jaspir Thapa at the Alexandra Hospital in February 1960.

Mediterranean populations was later found to occur all over the tropical world and to be one of the most common inherited diseases, and how later studies of its underlying cause helped to pave the way to the era of molecular medicine.

ADMIRAL OF THE FLEET THE EARL MOUNTBATTEN OF BURMA
KG, PC, GCB, OM, GCSI, GCIE, GCVO, DSO, FRS

BROADLANDS
ROMSEY.
HAMPSHIRE.
SO5 9ZD -

19th May, 1979

Dear Professor Weatherall,

Thank you so much for your extremely interesting letter of the 14th May. I showed it to my two daughters, Patricia Brabourne and Pamela Hicks who came to spend the weekend last night for the house opening today the 19th.

They and the rest of the family were all fascinated by what you had written and I shall place your letter in my wife's Archives which I have preserved in the main Archives here at Broadlands. I agree with you that there can be no doubt that you have identified this Gurkha child correctly.

As you probably know, my wife died on the 21st February in North Borneo so it is almost certain that this little child outlived her by a few days. The moving part about the picture is that she should have been cheering up a child who was dying and was herself so near death.

The only difference now is that we must record that the child was called Jaspir Thapa and was dying from thalassaemia.

An Indian artist did a very fine reproduction of the photograph which will hang in our Audio-Visual room at Broadlands, I return your transparancies, your medical paper and your cutting herewith.

It is indeed kind of you to have taken thetrouble to write and we are all fascinated about what you had to say. Thank you very much.

Yours sincerely
Mountbatten of Burma

2. Letter from Lord Mountbatten to the author in May 1979.

I

THE FIRST DESCRIPTIONS OF THALASSAEMIA

While many classical descriptions of disease followed astute observations at the bedside, others, particularly those affecting the blood, required advances in the laboratory sciences that underlie clinical practice. This is particularly well exemplified in the advances in haematology, the science of the blood, which led to the first description of thalassaemia in 1925.

Haematology at the Beginning of the Twentieth Century

It is generally agreed that the era of modern haematology stemmed from the work of the German pathologist Rudolf Virchow (1821–1902). Among his numerous contributions, Virchow developed specific staining systems with which to examine the different cells of the blood and to study those that contained a nucleus, the white blood cells, and those that were anucleate, the red blood cells. It was later discovered that the red blood cells are the product of nucleated precursors in the bone marrow that lose their nuclei during maturation.

Young red cells, called reticulocytes, still contain nuclear material that can be identified with certain stains. Normally, only 1–2 per cent of the red cells in the blood are reticulocytes: an increased level or the presence of nucleated red blood cells was found to indicate an increased rate of red-cell production. During the second half of the nineteenth century rather crude methods were developed for measuring the level of haemoglobin that is carried by the red blood cells, and hence for defining the severity of anaemia. Further important developments during this period included the invention of methods for measuring the numbers of red cells in the blood, as well as their size and haemoglobin content.

Towards the end of the nineteenth century and in the early years of the twentieth century Paul Ehrlich (1854–1915), developed improved techniques for staining blood cells. The importance of a careful examination of a stained blood film was developed further; indeed, it is still one of the most important haematological investigations. For example, it provided extremely important information about the cause of anaemia; the presence of pale-staining (hypochromic) or small (microcytic) red cells suggests a deficiency of haemoglobin, while large red cells (macrocytes) are found in disorders of defective red-cell maturation. Over the same period there was also an increasing interest in the physical properties of red cells, particularly their relative resistance to destruction, or lysis, when exposed to different concentrations of saline solutions.

The development of knowledge about the production and properties of the red cell during the first half of the twentieth century allowed a start to be made, albeit tentative, in the classification of the different forms of anaemia. While this work was progressing, clinicians became increasingly aware of diseases characterized by variable degrees of anaemia associated with considerable enlargement of the spleen, the splenic anaemias.

Splenic Anaemia

Unless we are unfortunate enough for it to be ruptured during an accident, most of us go through life without giving much thought to our spleen, a small organ hidden away below our ribs on the left side of our abdomen. As pointed out by the American haematologist William Crosby in his fascinating account of the evolution of knowledge about the spleen, except for a role in blood production during very early development, its function remained a complete mystery during the period of rapid development of haematology at the beginning of the twentieth century. However, if we are to appreciate the achievements of those who first described thalassaemia, it is helpful to understand the context in which an enlargement of the spleen first became associated with diseases of the blood.

The ancients assigned the spleen to the digestive system and often paired it with the liver. One of the first anatomists, Erasistratus, believed that, apart from maintaining the symmetry of organs in the upper abdomen, the spleen had no function. In the Hippocratic system of the four essential humours, blood, phlegm, yellow bile, and black bile, the spleen was the source of black bile. Later, Galen assigned a digestive function to the organ, which required that it had an outlet into the stomach. He suggested that humours unsuitable for its nutriment are discharged by the spleen through a canal into the stomach, one of several of Galen's anatomical arrangements that was later to be disproved by Vesalius. Early in the seventeenth century Christopher Wren, at that time working in Oxford, carried out one of the first recorded splenectomies, demonstrating that the removal of the spleen from a dog appeared to have no ill effects, a result that was replicated on many occasions over the next 200 years, not just on dogs but on a few humans. But, despite some progress in describing its anatomy, the functions of the spleen were to remain a complete enigma until the middle of the twentieth century. Crosby summarizes an account of the early developments of knowledge about the spleen by suggesting that twentieth-century haematologists inherited a single statement of value from the work of Galen: that the spleen is the *plenum mysterii organon*, the organ filled with mystery. It is embarrassing to admit that the phrase is still relevant to the twenty-first century!

The discovery that finally moved the spleen from the digestive to the blood system was based on the work of Virchow following his first description of leukaemia in 1858, a condition characterized by the abnormal proliferation of the white blood cells. He, and others, noticed that patients with more chronic forms of leukaemia have enlarged spleens. At first this disease was called splenic leukaemia. However, it gradually became clear that the association between anaemia and an enlarged spleen is not confined to leukaemia, and hence in 1866 Gretsel coined the term 'splenic anaemia' to cover these other entities. The term 'splenic anaemia' was popularized in the USA by William Osler, who, in a series of papers, described further cases and attempted to determine whether they represented a single disease. However, in a paper given before the Association of American Physicians in 1902 it was clear that, even in Osler's hands, the term 'splenic anemia' covered a completely heterogeneous series of clinical entities.

A particular form of splenic anaemia, restricted to early life and which became known as 'anaemia infantum pseudoleucaemica', was first described by Rudolf von Jaksch-Wartenhorst (1855–1947), who spent most of his professional life in Prague and was one of the first to appreciate the importance of chemical and radiological investigations in clinical practice. His original description of this disease, published in 1899, concerned a young boy with anaemia, a raised white-blood cell count, and enlargement of the spleen in whom a subsequent autopsy did not show changes of leukaemia. This condition, under the title of 'von Jaksch's anaemia', appeared frequently in the medical literature over the next twenty-five years, but on reviewing these clinical descriptions today it is clear that they dealt with an extremely heterogeneous collection of disease entities.

The First Descriptions of Thalassaemia

Although the credit for the first clinical description of what became known as thalassaemia is given to the American paediatrician Thomas B. Cooley (1871–1943) (Figure 3), it is quite clear that milder forms of the condition were also identified as a distinct entity at about the same time by several Italian clinicians. But undoubtedly Cooley was responsible for the first description of the severe, life-threatening form of the disease, while the Italian workers described milder varieties.

Cooley was a seminal figure in the evolution of paediatric haematology. After early training in Michigan and internships at Boston City Hospital, and a stint of service in the First World War, he spent the rest of his life in Michigan, first as a practising

3. Thomas B. Cooley.

paediatrician and later as Professor of Pediatrics in Detroit. Remarkably, he had no formal training in haematology and his equipment consisted of an ancient microscope, a rack for staining blood films, and a small card file, all housed in an otherwise empty room in which there was a couch on which he took regular siestas.

Cooley's first account of what became known as thalassaemia appeared in a single-page abstract in *Transactions of the American Pediatric Society* in 1925. Written together with his long-standing assistant, Pearl Lee, the abstract describes five young children with anaemia and enlargement of the spleen and liver, together with discolouration of the skin and curious mongoloid appearances of the skull (Figure 4). The authors were also very impressed by the remarkable uniformity of the abnormal radiological appearances of the skull and long bones in each of the children (Figure 5) and by the abnormal morphology of their red cells, which were pale and varied in

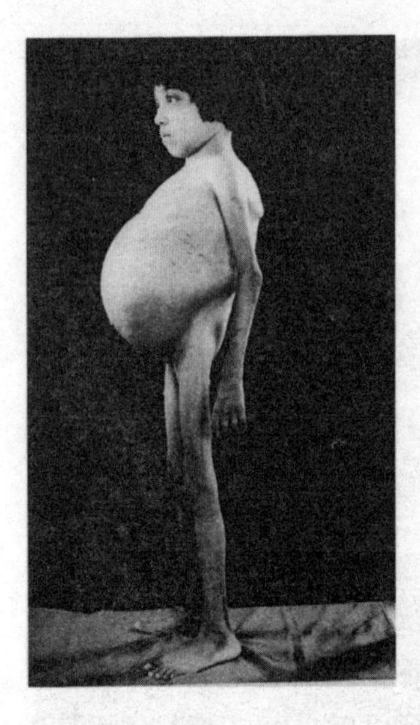

4. A child with severe thalassaemia. The figure shows the facial changes together with severe enlargement of the abdomen that is due to a massively enlarged spleen. The child also shows the pigmentation characteristic of the disease.

shape and size (Figure 6). They suggested that the radiological changes of the disease might be a valuable aid to diagnosis.

In this initial report three of the patients had died, but one, who had received a course of treatment for syphilis because of a weakly substantiated diagnosis of congenital syphilis, began to improve after cessation of treatment and at the time of writing appeared to be on the road to recovery. The fifth patient underwent splenectomy, which did not improve his condition. The authors noted that after the spleen had been removed there were enormous numbers of nucleated red cells in the child's peripheral blood. They describe how further treatment with a mixture of spleen and red bone marrow, combined with the administration of hydrochloric acid, had little therapeutic effect; the same lack of benefit was also observed after a single

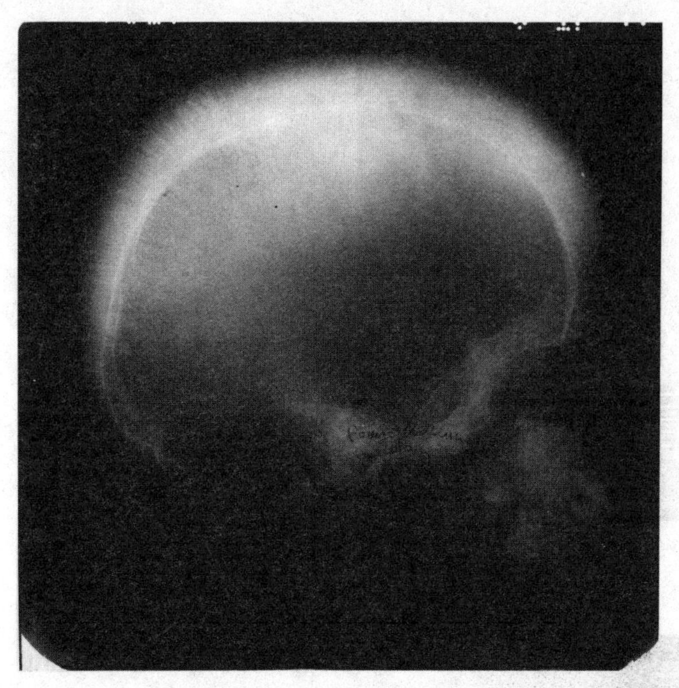

5. A skull X-ray from a child with severe thalassaemia. The typical 'hair-on-end' appearance that is due to expansion of the bone marrow is very clearly shown.

blood transfusion. Another notable feature of their report was a brief description of the microscopic appearances of the tissues, presumably at autopsy, which were described as showing aplasia of the red-cell-forming tissue, suggesting that, since these changes were present from very early in life, the disease might be some form of 'myelophthisic anaemia'. They also wrote that, in one of their cases, there was evidence that 'the body may compensate through secondary haematopoietic areas for the primary aplasia'.

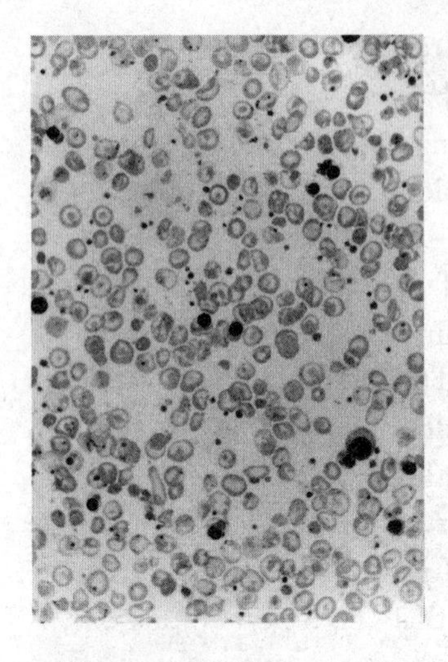

6. The peripheral blood film of a child with thalassaemia. The red cells show abnormal morphology with pallor, variation in shape and size, and many target cells characteristic of the condition. There are also many dark-staining nucleated red blood cells present, which are not normally seen in the peripheral blood.

The only puzzling feature in this otherwise remarkably accurate picture of children with severe forms of thalassaemia is the brief note on the autopsy findings that describes 'aplasia' of the red-cell-forming tissues. The term 'myelophthisic anaemia', which is now no longer used, appeared in the haematological literature until the middle of the twentieth century. For example, in the sixth edition of Muir's classical *Text-Book of Pathology*, published in 1951, it is described as an anaemia resulting from lesions of the bones involving the marrow, intimating that the primary cause of the disease resides in the bones. As we shall see in subsequent chapters, and as was clearly described in autopsy reports of thalassaemia only a few years after Cooley's early description of the disease, the red-cell-forming parts of the bone marrow are extremely active in this disease. It is surprising, therefore, that in the first report it is described as 'aplastic', while, at the same time, the authors clearly recognized the extensive areas of red-cell production in what they term the 'secondary haematopoietic areas', meaning the liver and spleen, which are not usually involved in blood-cell formation except in fetal life.

The great contribution of this initial description and subsequent papers was to define a series of children with a specific clinical syndrome and to separate it clearly from the heterogeneous group of childhood anaemias that hitherto had gone under the general title of splenic anaemia, or 'von Jaksch's anaemia'. Cooley was one of the first clinicians to appreciate the importance of genetics in medical practice, but, although he recognized that children with this condition were born with it, and thus that it is a congenital disease, he never seems to have considered that it might have had a genetic basis.

In the same year that Cooley's first description of thalassaemia appeared, Fernando Rietti described a patient in Ferrara with a milder form of anaemia associated with jaundice and enlargement of the spleen in which there were striking abnormalities of the red blood cells. He also observed that the red cells showed increased resistance to lysis in dilute saline solutions, a particularly important observation at the time, because it clearly distinguished the disorder from a form of familial anaemia that had been reported earlier called hereditary spherocytosis, in which the red blood cells show decreased osmotic resistance. Similar descriptions were published shortly afterwards by other Italian workers including Ezio Greppi and Ferdinando Micheli, and for many years this condition was referred to in the Italian literature as 'La malattia di Rietti–Greppi-Micheli'. In reviewing this early Italian literature, it is clear that the disease that was described under this name had many similarities to the more severe form described by Cooley, but that, unlike the latter, many of the patients survived into adult life. It is also clear that there were familial forms of the condition. The significance of this does not seem to have been recognized by the authors, but undoubtedly these early descriptions by the Italian workers reflected both the remarkable heterogeneity of thalassaemia and, probably, the first descriptions of its carrier states.

Why 'Thalassaemia'?

This brief outline of the early descriptions of thalassaemia raises a number of intriguing questions, not the least of which is how the disease received its unusual name. The term 'thalassemia' was first used by Whipple and Bradford in 1932 in a classical paper on the pathology of the disease. The word is taken from the Greek θαλασσα meaning 'the sea', reflecting the fact that all the early cases that were identified in the USA had a Mediterranean origin. The disease had already collected a number of names, including Cooley's anaemia, Mediterranean disease, Mediterranean anaemia, and so on, none of which Whipple and Bradford were happy with. In a delightful essay on the life and work of George Whipple (1878–1976), who was later awarded the Nobel Prize for his contribution towards an understanding of nutritional factors in anaemia, Lemuel Diggs makes it quite clear that it must have been Whipple and not Bradford who suggested the name. He describes Bradford as 'a competent baseball player who was quite familiar with the north end of south-bound mules and the ways of quail and pheasant but who did not know a word of Greek'. When asked by Diggs why he changed the name of the disease to 'thalassaemia', he replied: 'a good baseball player does not argue with the umpire.'

In fact, Maxwell Wintrobe and I heard the story of how thalassaemia received its name from George Corner, who was at Rochester at the time when Whipple and Bradford were doing their early work on thalassaemia. Since there must be few examples of a better account of how a disease received its name, Corner's letter is worth quoting.

> Because I was perhaps the most bookish of the young Rochester faculty, Dean Whipple made me his informal consultant on literary matters, several times asking my opinion on questions of nomenclature and etymology. Wishing to avoid the eponymic title 'Cooley's anemia', he sought a name that would associate the disease with the Mediterranean area, all the cases known at that time having occurred in families originating there. He had studied Greek at Phillips-Andover Academy; he recalled the great story in the Anabasis of Zenophon's army coming over the mountain and gazing at last upon the sea, the Ten Thousand shouting as one man, 'Thalassa, Thalassa!'. Whipple sent for me and asked whether I thought the name 'thalassic anemia' correct and appropriate. I had, in fact, never studied Greek, but of course I knew about the retreat of the Greek army from Persia and could at least tell the Dean that both words of his proposed name were from Greek roots and therefore properly associated. I gave no thought to the geographical aspects of the problem. Not until long afterwards did I learn that the view hailed so joyfully by the homeward-bound Greeks was actually the Black Sea. The weary men still had a voyage before them, and the Bosphorus to pass before reaching the Mediterranean Sea.

By the time that Corner made this embarrassing discovery, the name 'thalassaemia' was already well ensconced in the literature.

As it turned out, however, the minor classical deficiencies of the Rochester school were not to matter. Many years later it was found that the Black Sea is an equally suitable location to give its name to the disease, because it is surrounded with affected populations on all sides!

Why was the Severe Form of Thalassaemia First Described in the USA?

Another interesting question is why it was that the serious forms of thalassaemia were first recognized in the United States, where they were very uncommon, and not in Italy or other Mediterranean countries where they must have occurred at a very high frequency. The answer almost certainly lies in the pattern of disease in early childhood in the Mediterranean region in the early part of the twentieth century. At the turn of the century and up to the mid-1930s, in Italy alone there were over 200,000 cases of severe malaria notified each year, and between 2,000 and 10,000 deaths. In Greece during the same period the number of deaths attributed to malaria varied between 3,000 and 8,000 per annum, and there were similar mortality rates in the southern parts of Spain. Since malaria causes severe anaemia and enlargement of the spleen in early life, severe thalassaemia would not have been recognized as a distinct entity in the Mediterranean countries during the first part of the twentieth century. In addition, there were other common infections that would have produced a similar clinical picture, notably leishmaniasis. It is not surprising, therefore, that the disease was first recognized in the USA, where this extremely high frequency of infectious disease in childhood did not occur. The forms of the disease that were identified independently in Italy by Rietti and his colleagues were very much milder and compatible with survival to adult life.

It is also possible that the lack of recognition of severe forms of thalassaemia in Italy reflects to some degree the way in which haematology developed in the early part of the twentieth century; the study of the morphology of blood cells was most active in Pavia and Siena, where there would have been relatively little thalassaemia.

Earlier Descriptions of Disease that Resembled Thalassaemia

Is there any evidence that thalassaemia was recognized as an entity before 1925? A Greek physician, Caminopetros, noted one possible reference to a disease of this kind in Hippocrates' *Coan Prognosis*. Bannerman, in his short monograph on thalassaemia, quotes an English translation as follows: 'When children of 7 years of age show weakness, a bad colour and rapid respiration on walking, together with a desire to eat earth, it denotes destruction of the blood and asthenia.' It is much more likely, however, that this was a description of iron-deficiency anaemia in early childhood; a desire to eat dirt, a condition known as pica, occurs with iron deficiency and is not characteristic of thalassaemia. Menke made the more plausible suggestion that there is a description of sickle-cell thalassaemia, described in a later chapter, in a volume of the *Hippocratic Collections*, 'Concerning Internal Diseases', entry 32. He quotes a Latin version given by Ermerins as follows:

> Another disease of the spleen. It begins indeed in the time of Spring but more especially from the blood. For when the spleen is filled with blood it bursts out into the abdomen and sharp pains attack the spleen, the breast, the shoulder,

> and under the shoulder-blades, the entire body is of a leaden colour and over the shins are minor scratches from which originate large ulcers.

Although there are a number of other conditions that might present in this way, this is certainly compatible with sickle-cell thalassaemia, or sickle-cell anaemia, diseases that we will discuss in later chapters.

The Italian haematologist Ida Bianco has brought my attention to some early Italian literature published between 1880 and 1890 in which there are reports of children with splenic anaemia, in some cases with more than one affected family member, bone changes similar to thalassaemia, and generalized thinning of the bones, findings that are particularly difficult to reconcile with an infective origin to the disease. These early Italian authors, notably Cardarelli, were clearly puzzled by this condition, and, although they included various infections in their differential diagnosis, they also recognized that there might be constitutional factors involved.

An interesting approach to the question of whether thalassaemia occurred in historic or prehistoric times is the study of peculiar thalassaemia-like bone changes in skulls found in ancient burial places, or mummies in a good state of preservation. There is a condition well known to anthropologists called 'porotic hyperostosis' in which the structural and radiological changes in the skull are very similar to those of severe thalassaemia. Changes of this type have been found in skulls excavated in Sicily and Sardinia as well as those of the ancient native populations of America, the Incas of Peru, Indians from Colombia, Aztecs from Mexico, and Mayan Indians from Yucatan. Several palaeontologists believe that one contribution to the extinction of some of these ancient populations might have been a high incidence of a blood disease, possibly a genetic anaemia like thalassaemia. More recently similar skeletons have been discovered in burial grounds in central Thailand. While these could be the remains of patients with a disease like thalassaemia, there are other forms of anaemia that cause expansion of the bones of the skull.

One of the major arguments that was raised against the concept that porotic hyperostosis reflects severe thalassaemia was the observation that it had been observed mainly in adult skeletons; without treatment, children with severe thalassaemia died early in infancy. In written evidence to the Paleopathology Association's meeting in Detroit in 1977, I pointed out that the age of the skeletons did not exclude the diagnosis of thalassaemia, because the milder forms, which are also associated with these skull abnormalities, are compatible with survival into adult life. These discussions of course occurred before more recent evidence about the likely time of expansion of the genes for thalassaemia, a topic to which we will return in a later chapter. Until studies of the dried blood of these remains, or, better, their DNA settle this fascinating debate, the origin of these extraordinary skeletal changes will remain one of biology's many unsolved mysteries.

The First Ten Years

In the second edition of her remarkable monograph *The Anaemias*, published in 1936, Janet Vaughan summarized what had been learnt about thalassaemia in the first ten years after its recognition. It was viewed as a disease characterized by profound anaemia from early life, enlargement of the spleen, extraordinary mongoloid deformities of the face and skull, pigmentation of the skin, and characteristic radiological changes of the skull and long bones. Death usually occurred in early childhood, and there was no satisfactory form of treatment. Over half the section on thalassaemia in Vaughan's monograph is devoted to the findings at autopsy in this disease, based partly on the extensive publications of Whipple and Bradford, but also on a personal communication from one of her colleagues in London. These descriptions have never been bettered. In particular they cover the extensive accumulation of iron in many of the tissues, changes in the liver and heart consequent on iron loading, the enormous expansion of the red-cell-forming precursors in the bone marrow, and the thinning and curious structural changes of the bones. The widespread pigmentary changes of the skin are also emphasized.

Vaughan also mentions that in a few reports there had been multiple cases within families, and it was suggested that there might be a hereditary element. As regards the basic cause, the early suggestions of Whipple and Bradford are quoted, suggesting that the disease reflects 'some fundamental change in the method of handling body pigment'.

Interestingly, though well referenced for its time, Vaughan's account makes no mention of any work on the milder forms of the disease in Italy. This is not surprising, however. The early studies of Rietti and his colleagues on the milder forms of thalassaemia were all written in Italian and published in local journals, as were the descriptions of the severe forms of thalassaemia that were published by Italian workers in subsequent years. There were, of course, no international journals of haematology at the time and no international meetings in the field. Indeed, and as we shall see in subsequent chapters, it was only after the Second World War that there was genuine international recognition of the relative roles played by workers in Italy and Greece and those in northern Europe and the United States in the early development of this field.

While in retrospect it might appear that relatively little progress was made towards an understanding of thalassaemia during the first ten years after its discovery, it is important to remember that the extremely accurate descriptions of the clinical features and post-mortem changes of the disease formed a solid basis for the later understanding of its underlying mechanisms and inheritance. Furthermore, considering the extreme heterogeneity of diseases that were labelled 'splenic anaemia' during the early years of the twentieth century, the clear definition of a disorder of this kind as a distinct entity was no mean achievement.

II

THALASSAEMIA AS A GENETIC DISEASE

Although there had been sporadic reports of the occurrence of disease in more than one family member over several centuries, the concept of genetic disease was not born until the beginning of the twentieth century, following the rediscovery of the work of Mendel. In 1899 and 1901 the British clinician Archibald Garrod published accounts of a condition called alkaptonuria, characterized by joint disease and the passage of dark pigment in the urine. He pointed out that the disease occurs in siblings and, in the second paper, emphasized the frequency of consanguinity in their parents. Although it is clear that he did not appreciate the significance of these findings, William Bateson, a botanist and one of the leading protagonists of Mendel's work at the time, became aware of them and pointed out to Garrod that they were just what might be expected in a disease that had a recessive form of inheritance—that is, that it was passed on by each of the symptomless parents to the affected children. Garrod continued to study rare inherited disorders of this type and, in June 1908, presented this work in full in his Croonian Lectures to the Royal College of Physicians of London. Although these lectures, entitled 'Inborn Errors of Metabolism', marked the beginning of the new field of medical genetics, relatively little attention was paid to this work for many years. Indeed, genetics became an integral part of medical practice only after the Second World War.

Garrod's work set the scene for the analysis of genetic disease by the application of Mendel's Laws, which anticipated two main forms of inheritance, recessive and dominant. In the former, healthy parents who carry only one copy of a defective gene are called heterozygotes, or carriers. Each of their children has a one in four chance of inheriting the defective gene from both parents and hence of suffering from the inherited disease. They are then referred to as homozygotes. Unlike recessive inheritance, dominant inheritance requires the presence of only a single defective gene and is therefore passed on directly from generation to generation.

The First Genetic Studies of Thalassaemia

Although the occurrence of more than one family member with thalassaemia was mentioned several times in the early literature, the significance of this observation was not discussed further. The first suggestion that it might be an inherited disease was made in 1932 by Whipple and Bradford in their paper that gave the disease its name; indeed it is implicit in the title of the paper, 'Racial or Familial Anemia of Children'. The suggestion that it could have a genetic basis appears in one sentence in the summary. A similar suggestion was made by Cooley and Lee in the same year, and in 1934 Moncrieff and Whitby postulated that it might be inherited in a Mendelian fashion.

It was not until 1937, twelve years after Cooley had first described the severe form of thalassaemia, that it was clearly shown to be a genetic disease. Between 1935 and 1938 work by Micheli and Angelini in Italy and Caminopetros in Greece provided the first evidence about the mode of genetic transmission of thalassaemia. These workers noticed that parents of patients with the severe form of thalassaemia, though not anaemic themselves, had red cells with increased osmotic resistance—that is, they did not lyse as readily as normal red cells in dilute saline solutions. The work of Caminopetros was of particular importance, although, as pointed out in a later appreciation by Phaedon Fessas, a Greek haematologist and major figure in the thalassaemia field, the results of his family studies were rather strange, because he usually found increased osmotic resistance in only one parent of a child with severe thalassaemia. However, he was extremely explicit in his conclusions:

> the increased osmotic resistance is found constantly among seemingly healthy parents and siblings of the deceased. These findings speak for accepting the hereditary transmission of the pathologic element. The only possible explanation is a lesion of haemopoiesis, which is hereditary and idiopathic. Therefore, the disease must be considered as being transmitted by heredity, in fact as a recessive carrier, according to the laws of Mendel, as can be concluded by the transmission through seemingly healthy individuals.

These studies were extended in Italy, notably by Ezio Silvestroni and Ida Bianco, workers who were also to make extensive contributions to thalassaemia research for many years. Again, using the osmotic fragility test they noted that the parents of children with severe forms of thalassaemia both had red cells with reduced osmotic fragility and they went on to use this simple test to analyse the frequency of the disease in different parts of Italy. However, like Caminopetros, they noted that in some families only one parent was affected, usually in cases where the form of thalassaemia in the affected children was milder.

At the same time as these early genetic studies were being carried out in Greece and Italy, similar observations were being made quite independently in the USA. In 1940, Max Wintrobe and his colleagues at Johns Hopkins Hospital, Baltimore, described small pale red cells with increased osmotic resistance in some of the relatives of an Italian patient with moderately severe thalassaemia. They realized that these findings in the relatives probably reflected a mild form of thalassaemia and, in a footnote to their paper describing these observations, they pointed out that they had seen this haematological picture in both parents of a child with severe thalassaemia. In later accounts of this work they gave credit to the astuteness of the laboratory technician who was helping them to examine these families, Regina Weistock, for noticing these subtle red-cell changes.

Similar observations were published independently at about the same time by several other American haematologists, including William Dameshek, one of the major figures in the development of haematology in the USA.

Further extensive family studies were carried out in the USA, notably by William Valentine and James Neel, whose work on the genetic transmission of thalassaemia was seminal. Neel, who later became an extremely distinguished human geneticist, wrote many years later in his autobiography about how he was stimulated to carry out these important studies by reading Wintrobe's account of the Baltimore family.

By the end of the 1940s, and following the beginnings of better scientific communication between Europe and the USA, it was possible for the results of these different studies to be more fully assessed and reviewed. By then it was clear that the severe anaemia described by Cooley is the homozygous state for a recessive Mendelian gene, and that the heterozygous, or carrier, state is characterized by extremely mild anaemia with osmotically resistant red cells. For many years a different nomenclature for the carrier state was used in Italy and the USA; in Italy it was called 'thalassaemia minima' or 'microcytemia', while in English-speaking countries Valentine and Neel's 'thalassaemia minor' was favoured.

Why is Thalassaemia so Common in Mediterranean Populations?

While carrying out these early studies on the genetic transmission of thalassaemia, workers on both sides of the Atlantic were puzzled as to why there were remarkably high carrier rates for the disease in Mediterranean populations, while populations of north European origin on both sides of the Atlantic did not seem to be affected. During this period sporadic reports of the occurrence of the major form of the disease started to appear from other parts of the world, but it was only later that its remarkable high frequency throughout the Middle East and many parts of Asia was discovered.

In the period immediately after the Second World War, and after the atomic bombs had been dropped on Hiroshima and Nagasaki, there was considerable concern about mutation rates in human beings and how these might be modified in the Japanese populations who had been exposed to radioactive fall-out. There were many theoretical studies directed at attempting to determine the mutation rate—that is, the frequency of mutations per generation at a particular gene locus. Extrapolating these ideas to the high frequency of thalassaemia, in 1948 Neel and Valentine reviewed their population studies in the USA. They assumed that, because thalassaemia homozygotes die early in life, their genetic fitness—that is, the ability of those with two mutant versions of the gene to pass on their genes—is zero. They also assumed that heterozygotes, who carry only one mutant form of the gene, do not come under a selective advantage. Basing their calculations on these uncertainties, they proposed a mutation rate for thalassaemia of 1:2,500. An even higher rate was proposed by Italian workers at the time, who also suggested that some form of positive selection might have operated to maintain the frequency of thalassaemia heterozygotes. Neel and Valentine even proposed that, to explain the difference in frequencies between different ethnic groups, there might be inter-ethnic variation in mutation rates.

These issues were discussed widely at the Eighth International Congress of Genetics in Stockholm in 1948, a meeting that was attended by J. B. S. Haldane (1892–1964). Haldane, who was always known as J.B.S. to distinguish him from his father, the famous respiratory physiologist J. S. Haldane, was one of the most remarkable scientific polymaths of the twentieth century. As early as 1923 and aged 31 years, in a lecture given to the Heretics, a Cambridge society, entitled 'Daedalus or Science and the Future', he predicted with remarkable accuracy how developments in the biological sciences would shape our futures, including the birth of the first test-tube baby. It was the publication of this essay that was to be the inspiration for Aldous Huxley's novel *Brave New World*. Later, together with the Cambridge statistician and geneticist Ronald Fisher, he developed many of the principles of quantitative genetics that still form the basis for the field today.

Haldane was not impressed with Neel's and Valentine's estimates of human mutation rates, particularly as they related to the frequency of thalassaemia. In the *Proceedings of the Congress* he wrote:

> Neel and Valentine believe that the thalassaemia heterozygote is less fit than normal, and think that the mutation rate is above 4×10^{-4} rather than below it. I believe that the possibility that the heterozygote is fitter than normal must be seriously considered. Such increased fitness is found in the case of several lethal and sub-lethal genes in *Drosophila* and *Zea*. A possible mechanism is as follows. The corpuscles of the anaemic heterozygotes are smaller than normal, and more resistant to hypotonic solutions. It is at least conceivable that they are also more resistant to attacks by the sporozoa which causes malaria, a disease prevalent in Italy, Sicily, and Greece, where the gene is frequent.
>
> Until more is known about the physiology of this gene in various environments I doubt if we can accept the hypothesis that it arises very frequently by mutation in a small section of the human species.

So was born the 'malaria hypothesis'. Curiously, in a review written many years later about Haldane's contributions, the American geneticist and Nobel Laureate Joshua Lederberg suggested that the concept of genetic resistance to infection was already well known by the time of Haldane's lecture and subsequent publication and hinted that he may have developed the 'malaria hypothesis' based on previous work by the Italian geneticist Montalenti. However, he does not refer to the paper quoted above, but discusses a paper based on a lecture given by

Haldane at the Symposium on Ecological and Genetic Factors in Speciation among Animals, held in Milan, the Proceedings of which, like those of the Stockholm Congress, were published in 1949. In fact, this lecture did not deal with malaria or thalassaemia at all. In the footnotes to the published discussion of the meeting Montalenti acknowledges a verbal communication from Haldane suggesting that thalassaemia carriers may be more resistant to malaria; in the same discussion Haldane added that they might also be at an advantage in iron-deficient environments. Lederberg also intimates that Haldane may have heard about the relationship between the distribution of thalassaemia and malaria in Italy from a prolonged correspondence with Montalenti and other Italian workers including Silvestroni who had studied the distribution of thalassaemia in Italy, work which they published in *Nature* in 1950 and which, incidentally, does not mention malaria at all but simply suggests that thalassaemia carriers may be more fertile.

In a paper also published many years later, Allison makes a similar suggestion, claiming that Haldane repeated Montalenti's proposal with no acknowledgement in his publications or lectures. As we have seen, however, Montalenti's paper published in 1950 does not suggest that he or his colleagues were thinking along these lines. In fact, several Italian workers had noted the relationship between the distribution of thalassaemia and malaria as early as 1946, when Vizzoso published the first paper to this effect. But none of them suggested that thalassaemia might be a malaria-protective condition. It is particularly interesting that neither Lederberg nor Allison quotes Haldane's paper from the *Proceedings of the Stockholm Congress,* which suggests that the thalassaemia trait may be protective against malaria. Some years ago I had the opportunity of discussing these issues with Ida Bianco, the wife and co-worker of Silvestroni, and it is clear that, although there was considerable interest in the relationship between the geographical distribution of thalassaemia and malaria in Italy, nobody was thinking along the lines expressed by Haldane at the Stockholm meeting, which, incidentally, occurred before the meeting at which Montalenti and Haldane met in Milan.

It seems clear, therefore, that Haldane's lecture in Stockholm was the first formal statement about thalassaemia and malaria resistance. Curiously, throughout the scientific literature Haldane's hypothesis is usually referenced to the proceedings of the Milan meeting; it is possible, therefore, that some of his detractors, who like Lederberg do not cite the Stockholm reference, were unaware that the meeting took place! Readers might wonder why we have delved into the origins of the malaria hypothesis in such detail. The fact is that it was to have major implications for future studies in human evolutionary biology; moreover, it was, arguably, the finding that arose from the thalassaemia and abnormal haemoglobin field that was to have the most important implications for the better understanding of human biology in general. Paradoxically, when, in 1954, the first intimations that Haldane was correct appeared, they were not in the thalassaemia field, but from related studies in Africa by Allison on the genetics of sickle-cell anaemia, a story to which we will return in a later chapter.

Haldane's remarkable insight, which only later became known as the 'malaria hypothesis', was the beginning of the concept that common genetic diseases like thalassaemia might reflect a state of balanced polymorphism, in which the protective effect against a disease like malaria for heterozygotes would allow the frequency of the gene to expand until it was balanced by the loss of severely affected homozygotes from the population. It was to take many years before it was possible to confirm or refute whether Haldane's remarkable prediction was correct in the case of thalassaemia, a theme to which we will also return in a later chapter.

Is there More than One Form of Thalassaemia?

In the late 1940s and after, several reviews in English were published by Italian workers in early volumes of new international haematology journals that leave no doubt that they were extremely puzzled by the clinical variability of thalassaemia. Because of their lack of knowledge of the nature of the disease, they were unable to provide any explanation for its remarkable clinical diversity and they simply had to postulate the existence of 'modifying factors' that might explain the wide range of clinical disorders that were being observed. The diseases that they described ranged from profound anaemia in early infancy, through milder anaemia with enlarged spleens in adults, to completely symptomless conditions identified only by the morphological or osmotic properties of the red cells. Because of this extraordinary heterogeneity of what appeared to be the same disease, the Italians Chini and Valeri, authors of an extensive review published in 1949, coined the term 'Mediterranean hemopathic syndromes' to describe these diverse disorders.

Hence, by 1950 it was clear that thalassaemia is a disorder that is inherited according to Mendel's laws and that occurs at a very high frequency in certain ethnic groups. It was also clear that within this general framework the disease exhibits remarkable clinical heterogeneity. But virtually no progress had been made towards an understanding of the underlying cause of the disease, in particular whether it is primarily a disorder of red-blood-cell production or whether the characteristic anaemia is a part of a much more general disturbance of metabolism, as hinted by Thomas Cooley and others in their early descriptions of thalassaemia.

III

THALASSAEMIA AS A GENETIC DISORDER OF HAEMOGLOBIN PRODUCTION

The remarkable successes in defining thalassaemia as a disorder of haemoglobin production that were achieved between 1950 and 1960 are an example of one of those unplanned episodes in the rapid advance of a field of science that reflects the chance coming-together of information from several different disciplines at the appropriate time. It is important to emphasize that such multidisciplinary success would never have been achieved so rapidly without the vast improvements in scientific communication that evolved after the Second World War with the publication of international journals and, of particular importance in the haemoglobin field, the evolution of major international interdisciplinary conferences at which, as well as the delivery of research papers, there were long periods of discussion, which were published after the meetings. The latter are of enormous value when trying to thread together the disparate thoughts of research workers in a field that was moving so rapidly.

Although the speed of development of the field during this relatively short time makes for considerable difficulty in putting events into a temporal setting, three particularly important sets of developments can be identified as the threads that came together to form the background for a synthesis that set the scene for a better understanding of the nature of thalassaemia. First, there was the discovery of the abnormal haemoglobins. At the same time it became clear that normal human haemoglobin is heterogeneous and that there are different haemoglobins at various stages of development. The third and equally important thread in the story was the remarkable progress that was made towards an understanding of the structure and genetic control of human haemoglobin. With this new knowledge, and the discovery of patients who had inherited both thalassaemia and an abnormal haemoglobin, and by analysing their patterns of haemoglobin production, it was possible to produce at least a tentative synthesis of the concept that thalassaemia is a disease of haemoglobin production.

The Discovery of the Abnormal Haemoglobins

The discovery of the abnormal haemoglobins occurred over much the same period as the first descriptions of thalassaemia. In 1910, James Herrick, a physician from Chicago, presented a paper to the Association of American Physicians in Washington DC in which he described a young student from Granada, West Indies, with chronic anaemia and curiously elongated and sickle-shaped red blood cells. Apparently, the first description of sickle-cell anaemia was received with very limited interest by the distinguished audience of American physicians. Over subsequent years further cases were described, and it became clear that the disease is largely though not entirely restricted to those of African origin. Like thalassaemia, it took many years for its pattern of inheritance to be determined, but in the late 1940s extensive studies in both the USA and Africa demonstrated that it is inherited as a Mendelian recessive disorder in exactly the same way as thalassaemia. It is clear that Herrick himself did not think that his discovery was of great importance; it is barely mentioned in his autobiography, published in 1949. Perhaps this is not surprising, because he is also remembered as the first to describe the clinical picture and significance of coronary thrombosis!

The discovery of the cause of sickle-cell anaemia followed a chance conversation in 1945 on an overnight train journey from Denver to Chicago between Linus Pauling, the distinguished chemist and Nobel laureate, and William Castle, the Boston haematologist who, among many other major contributions, is best known for his work in determining the cause of pernicious anaemia. Castle mentioned to Pauling that he had been examining the blood of patients with sickle-cell disease and had noticed that, when the red blood cells were deprived of oxygen and formed a sickle shape, they showed unusual properties when examined under polarized light. Pauling realized immediately that this might reflect some form of molecular reorganization and therefore that it might indicate that the haemoglobin of patients with sickle-cell disease could be in some way abnormal.

At about this time the Swedish chemist Arne Tiselius had developed a technique called electrophoresis, which made it possible to separate proteins of different structure and charge in an electric field, a discovery for which he received a Nobel Prize in 1948. Returning to his laboratory at the California Institute of Technology, Pauling, together with an extremely talented post-doctoral student, Harvey Itano, set about building a Tiselius apparatus and examined the haemoglobin of the red cells of patients with sickle-cell anaemia. They found that it moved in an electric field at a different rate from the haemoglobin of normal individuals and, even more importantly, that the red cells of unaffected carriers of the sickle-cell gene contained both normal and abnormal haemoglobin. This work was published in *Science* in November 1949, under the title 'Sickle Cell Anemia, a Molecular Disease'.

Although the Tiselius apparatus was too complicated for use in routine hospital laboratories, within a few years a much simpler method for searching for haemoglobin variants was developed using filter paper with a small laboratory power pack to supply the current. Soon laboratories all over the world were using this technique, and new abnormal haemoglobins were reported at regular intervals (Table 1). At first, they were designated by letters of the alphabet; normal haemoglobin was called haemoglobin A (Hb A), sickle haemoglobin was called haemoglobin S (Hb S), and numerous others followed (Figure 7). Indeed, the alphabet was soon used up describing these new variants, which were later named by the place in which they were discovered.

During the 1950s it was already apparent that, while haemoglobins S, C, and E occur at high frequencies in particular populations, most of the human haemoglobin variants are quite rare. A rich new field for human genetic investigation had been opened up, which was shortly to have very important implications for the further study of thalassaemia.

The Heterogeneity of Normal Human Haemoglobin

It had been known since 1866 that human placental blood, which of course contains large numbers of fetal red cells, is

TABLE 1 Some important human haemoglobins and their variants

This list does not include the embryonic haemoglobins and focuses only on those that were important in the early studies of haemoglobin genetics and thalassaemia.

Haemoglobin	Globin chains	Properties
Hb A	$\alpha_2\beta_2$	Main oxygen transporter from the first few months after birth.
Hb A_2	$\alpha_2\delta_2$	Minor adult haemoglobin. Elevated in β thalassaemia carriers.
Hb F	$\alpha_2\gamma_2$	Main haemoglobin during fetal development up to birth. Has a higher affinity for oxygen than Hb A.
Hb S	$\alpha_2\beta_2^S$	Causes sickle-cell anaemia in homozygotes.
Hb C	$\alpha_2\beta_2^C$	Causes a mild anaemia in homozygotes.
Hb E	$\alpha_2\beta_2^E$	Ineffectively synthesized and causes a mild form of β thalassaemia in homozygotes and a variably severe anaemia when inherited together with β thalassaemia—a condition called Hb E β thalassaemia.
Hb Hopkins 2	$\alpha_2^{Hop2}\beta_2$	Harmless α chain variant.
Hb I	$\alpha_2^I\beta_2$	Harmless α chain variant.
Hb H	β_4	Produced owing to a deficiency of α chains in α thalassaemia. It is unstable and has a very high oxygen affinity.
Hb Bart's	γ_4	Produced owing to a deficiency of α chains in fetal life and the neonatal period. Has a high affinity for oxygen.

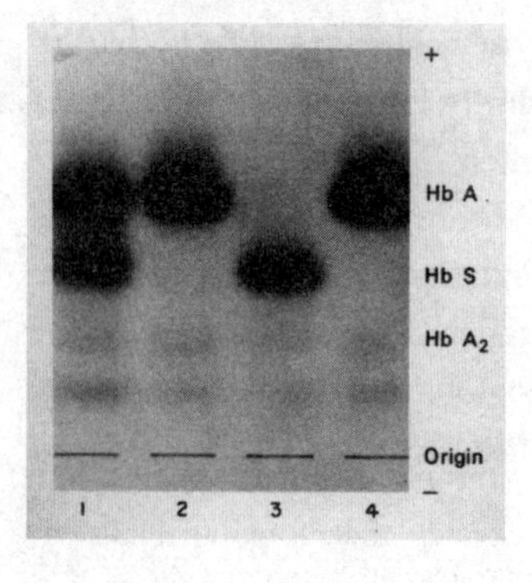

7. Haemoglobin electrophoresis on starch gel. The + and – signs reflect the orientation of the electric field. The positions of Hbs A, S and A_2 are shown. The dark band that migrates more slowly than Hb A_2 is a non-haemoglobin protein of the red cell. Track 1 shows the sickle-cell tract, tracks 2 and 4 are normal, and track 3 shows sickle-cell disease.

more resistant to denaturation by alkali than that of adults, a finding that was confirmed by newer techniques that were developed for studying proteins in the period running up to the discovery of Hb S. Clearly, therefore, human haemoglobin is heterogeneous; the predominant haemoglobin in fetal life is fetal haemoglobin, which was called haemoglobin F (Hb F), the level of which rapidly declines after birth and is replaced by adult haemoglobin, or Hb A. In the early 1950s methods were developed for studying haemoglobins by electrophoresis, which allowed larger samples to be applied than had been previously possible using filter paper electrophoresis or the Tiselius apparatus. In 1955, using these new approaches, Henry Kunkel and his colleagues in New York described a minor haemoglobin, present in all normal adults at a level of approximately 2.5 per cent of the total haemoglobin, which they called haemoglobin A_2. In the early 1960s it was found that human embryos have a different set of haemoglobins. Hence it became clear that human haemoglobin varies in its structure at each specific stage of development, presumably as an adaptive process to different oxygen requirements (see Table 1). Haemoglobin F has a higher affinity for oxygen than haemoglobin A and hence is more suited to oxygen delivery in the fetal environment and for the more effective extraction of oxygen from the placental circulation for the benefit of the fetus. The mechanisms that underlie the regulation of the switch from fetal to adult haemoglobin have remained elusive, a topic to which we will return in a later chapter.

By the mid-1950s, therefore, it was clear that there must be separate genes involved in the regulation of fetal and adult haemoglobin synthesis and, because a variant of Hb A_2 was described, a third set of genes must be involved in the regulation of the minor component of normal adult haemoglobin. The situation was further complicated when, in 1958, a team from Johns Hopkins Hospital described a large Baltimore family in which there were two abnormal adult haemoglobins, one of which was Hb S and another which they named 'hemoglobin Hopkins-2'. It was clear from the way in which these variants segregated within this large pedigree that they must be due to mutations at different gene loci and hence that there must be at least two genes involved in the regulation of adult haemoglobin. The fact that it was possible to make sense of this complicated system almost as soon as it was recognized is testament to the enormous progress that was being made in working out the structure of haemoglobin at the same time.

The Structure and Genetic Control of Human Haemoglobin

During the early part of the twentieth century there were remarkable technical advances in the study of protein structure that made it possible to determine the sequence of amino acids in the peptide chains of haemoglobin and also to obtain a detailed picture of the three-dimensional structure of the haemoglobin molecule by X-ray crystallography. As well as Rhinesmith, Schroeder, Huisman, Hill, Konigsberg, Itano, and Pauling in the USA, and Braunitzer in Germany, there are two other distinguished protein chemists, who had left Germany and moved to England before the Second World War, Perutz and Ingram, who, because of their particular interest in bridging the gap between protein chemistry and its application to research in human genetics, made major contributions to the field at this critical period of its development and later. And we should not forget the important contributions of another refugee to England from Germany, Herman Lehmann, a clinical chemist whose long-term scientific association with Perutz did so much to maintain the latter's interest in the clinical aspects of haemoglobin biology.

Max Perutz (1914–2002) was born in Reichenau, near Vienna, and in 1936 moved to the Cavendish Laboratory in Cambridge, where, apart from a period of internment in Canada as an 'enemy alien', he was to spend the rest of his career, later becoming head of the Molecular Biology Unit of the Medical Research Council and, from 1962 onwards, Chairman of the Medical Research Council's new Laboratory of Molecular Biology. Between 1937 and 1959, applying the methods of X-ray diffraction to the study of haemoglobin, and after innumerable setbacks, he finally defined the precise structure of the molecule, work for which he was later awarded the Nobel Prize. Shortly after he arrived in Cambridge in 1936, he met Hermann Lehmann. The two of them later shared the pleasures of internment as enemy aliens at Huyton, near Liverpool, though Lehmann was not deported to Canada and was appointed as a pathologist to the Royal Army Medical Corps. After the Second World War he worked in Uganda, where he first came across sickle-cell anaemia. After returning to England, he renewed his friendship and collaboration with Perutz, which continued for many years and which was one of the major factors in evolving an understanding of the relationship between the structure of abnormal haemoglobins and their defects in stability and function, work that was to have important implications for a better understanding of some forms of thalassaemia.

Vernon Ingram (1924–2008) was born in Breslau. After moving with his family to London, he obtained his early training at Birkbeck College, and, after postgraduate training in the USA, in 1952 he returned to England to join Max Perutz and Francis Crick at the Medical Research Council Unit at the Cavendish Laboratory, Cambridge. There he carried out his seminal studies, which identified the molecular structure of sickle-cell haemoglobin, and he became interested in the molecular pathology of thalassaemia. While in Cambridge he worked with several graduate students, notably Anthony Stretton and John Hunt, who were also to make valuable contributions to the haemoglobin field. Later, he moved to the Massachusetts Institute of Technology, where he spent the rest of his career in the pursuit of other areas of molecular biology.

It is fascinating to reflect that these extremely distinguished protein chemists and others provided the basis for the future development of the abnormal haemoglobin and thalassaemia fields only up to the early 1960s, after which most of them moved into completely different research areas of the rapidly expanding field of molecular biology. But there is no doubt that during this relatively short period their work laid the basis for remarkable developments, in the better understanding both of the relationships between the structure and function of the abnormal haemoglobins and of the molecular pathology of the thalassaemias.

It had long been apparent that haemoglobin must consist of complex ring structures called 'haem', which bind oxygen, combined with a protein called 'globin'. In 1956 Vernon Ingram found that haemoglobin consists of two identical half-molecules. These findings agreed beautifully with the X-ray crystallographic evidence of the three-dimensional structure of haemoglobin that had been amassed by Max Perutz and his colleagues over many years. Both approaches suggested that each half-molecule of globin is made up of two different peptide chains, which were named α and β, each of which carries a haem ring. Hence its structure can be represented as $\alpha_2 \beta_2$. At the same time as these studies were being carried out in England, the workers in Germany and the USA who were alluded to earlier established that human fetal haemoglobin, although having a grossly similar structure to adult haemoglobin, has a different sub-unit composition. One pair of peptide chains is identical to the α chains of HbA; the others, which are quite unlike β chains, were called γ chains. Haemoglobin F was represented, therefore, as $\alpha_2 \gamma_2$. Soon afterwards, it was found that Hb A_2 contains a fourth type of globin chain, δ, similar in structure to β, and again combined with α chains—that is, $\alpha_2 \delta_2$.

At the same time as these structural studies were being undertaken, further evidence about the genes that must be involved in the regulation of haemoglobin production was being obtained. Another critical discovery was made by

Ingram in 1956. He purified globin from the red cells of patients with sickle-cell anaemia, digested it into smaller peptide fragments with the enzyme trypsin, and separated the resulting digests in two dimensions, a method based on Frederick Sanger's previous work on insulin and called, at the time, fingerprinting (not to be confused with DNA fingerprinting, which came much later). Remarkably, the fingerprints of Hb A and Hb S were identical except for one peptide, which, when analysed for its constituent amino acids, showed only one amino-acid difference between normal adult haemoglobin and Hb S: glutamic acid in normal haemoglobin is replaced by valine in sickle-cell haemoglobin. This seminal discovery, which never received the level of recognition that it deserved, showed that the basis for the sickle-cell mutation is a single amino-acid substitution and also that the product of a particular gene must be a single peptide chain. In 1958 the Americans Beedle and Tatum received the Nobel Prize for their earlier work demonstrating that there is a direct relationship between a gene and an enzyme, the one-gene one-enzyme principle; now Ingram had extended this concept to the one-gene one-peptide-chain level. In short, Hb S results from a single amino-acid substitution in the β chain of haemoglobin, reflecting a mutation of the β globin gene.

Further studies by Itano and Robinson of the remarkable Baltimore family described earlier demonstrated that Hb Hopkins 2 is in fact an α chain variant, indicating that there must be separate pairs of genes for both the α and β globin chains and that, because of the patterns of inheritance of the α and β chain variants in this family, these genes are likely to be on different chromosomes. Hence by 1959 it was clear that the regulation of haemoglobin would require at least four sets of gene loci, α, β, γ, and δ. In the early 1960s the demonstration that there were two further different peptide chains in embryonic haemoglobins added two more genes to the list. Thus, the stage was set to make some sense of some of the early findings in patients with thalassaemia and those who had received a thalassaemia gene from one parent and a gene for a haemoglobin variant from the other.

Since these complex interactions were so crucial to the further understanding of the thalassaemias, and formed the basis for a better appreciation of their complex genetics, they are discussed in more detail in the section that follows. Readers who are unfamiliar with the principles of genetics may find it helpful to refer to Figures 8 and 9, which show a simplified version of these interactions.

The Haemoglobin Constitution of Patients with Thalassaemia

In the period running up to and during the seminal work described in the previous sections, several valuable clues appeared, leading to the concept that thalassaemia is a genetic disorder of globin chain production. As early as 1946 it was noted that the haemoglobin of patients with severe thalassaemia was more alkali-resistant than normal adult haemoglobin, suggesting that these patients have more fetal haemoglobin than is usually found after the first year of life. In 1952 Arnold Rich suggested that thalassaemia might result from a defect in Hb A production with persistent production of Hb F. The finding of elevated levels of fetal haemoglobin in many children with severe thalassaemia was replicated by many workers over this period.

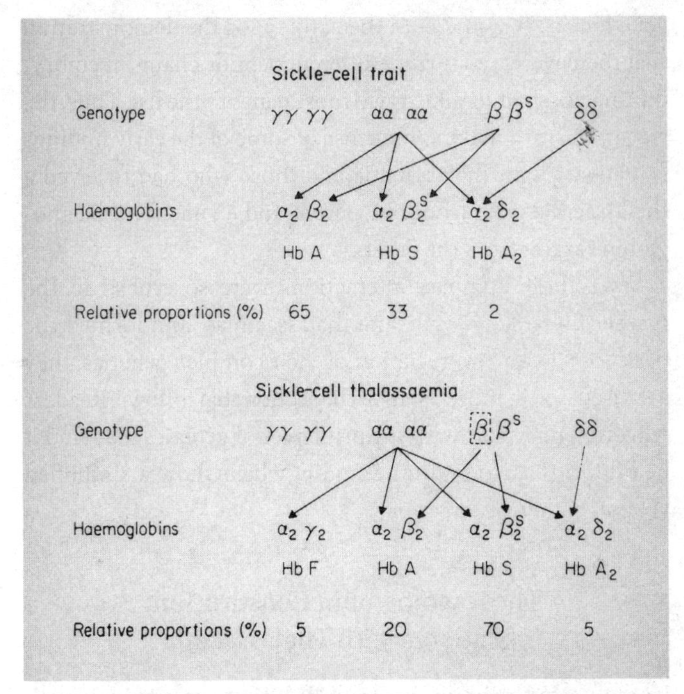

8. The haemoglobin patterns in the sickle-cell trait and sickle-cell β thalassaemia. The different globin genes are shown across the top of each condition. Because the β chains of Hb S are produced less effectively than normal, those with the sickle-cell trait have less Hb S than Hb A. However, those who have inherited both the sickle-cell gene on one chromosome and β thalassaemia on the other (shown by the dotted box), the situation is reversed, because the action of the β thalassaemia gene is markedly to reduce the amount of β chains; therefore, these patients have considerably more Hb S than Hb A.

In 1957, two years after the discovery of the minor fraction of normal human haemoglobin, Hb A_2, Henry Kunkel's team in New York observed that most but not all parents of children with severe thalassaemia had elevated levels of this

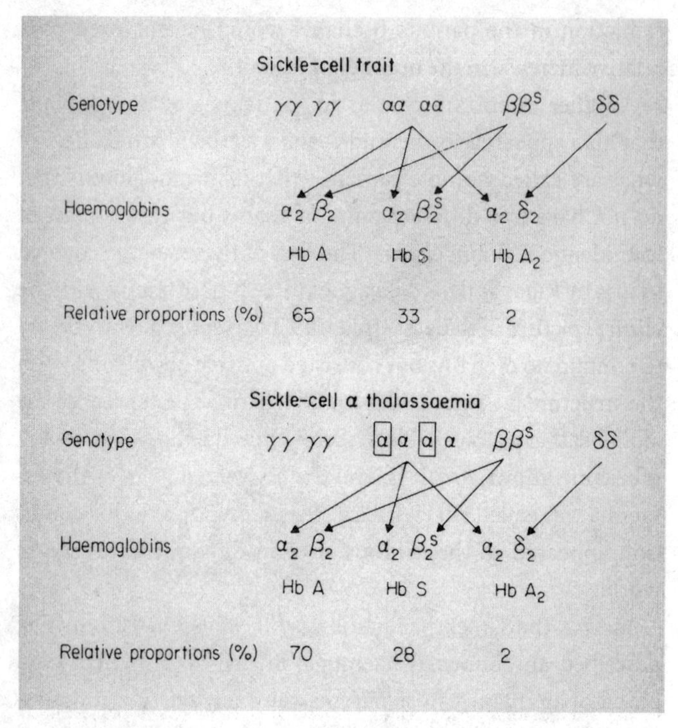

9. Haemoglobin changes in the sickle-cell trait and in those who have the sickle-cell trait and α thalassaemia. Unlike the situation in Figure 8, in this case defective α chain production affects both Hb A and Hb S equally, and so patients with this combination of genes show similar proportions of Hb A and Hb S, as seen in those with the Hb S trait alone.

component, usually in the range of 4–6 per cent of the total haemoglobin, about twice the normal level. In the event, this finding turned out to be of extreme value in the diagnosis of many common forms of thalassaemia, while, at the same time, it added further credence to the concept of one form of thalassaemia as a disorder of β globin production; a mild reduction in the parents β chains would be mirrored by a relative increase in the number of δ chains.

Another important clue as to the nature of thalassaemia that also appeared in the mid-1950s was the identification of what are called homotetramers—that is, haemoglobins that do not have two different pairs of chains but that consist of four identical globin chains. The first of these was discovered in 1955 by Rigas and his colleagues in a Chinese family with the clinical picture of thalassaemia, and it was subsequently found to contain no α chains but consisted entirely of β chains, with the structure β_4. Since it was found when the next letter of the alphabet that was available to name a new haemoglobin was H, it became known as Hb H, and the associated form of thalassaemia was called Hb H disease. Several reports of this condition appeared in the mid-1950s from different parts of the world.

In 1957 the Greek haematologists Fessas and Papaspyrou described an abnormal haemoglobin in the umbilical cord blood of an infant whose parents showed evidence of thalassaemia. Pending further identification, it was called Hb 'F and P'. A similar variant was discovered by Ager and Lehmann at St Bartholomew's Hospital, London, in 1958, and, because there were no letters of the alphabet left to name new haemoglobins, it was called Hb Bart's. It was soon found that Hbs 'F and P' and Bart's are one and the same and that they consist of four γ chains, with the structure γ_4. They are, in fact, the fetal counterparts of Hb H. There was considerable speculation at the time that these haemoglobin variants with no α chains might reflect a genetic defect in α chain production.

Undoubtedly the most important observation of all, leading to the concept that the thalassaemias result from reduced globin chain production, was the result of studies of the haemoglobin patterns of patients who had inherited the sickle-cell gene from one parent and a thalassaemia gene from the other, a condition called sickle-cell thalassaemia. This disorder was first described by Silvestroni and Bianco in Italy in a series of papers published between 1944 and 1945, in which they called the condition 'microdrepanocytic disease'. At that time it would have been too early for the Italians to recognize the full significance of their discovery, and it was not until 1952 that American workers analysed the haemoglobin patterns of patients with this condition. They found that the red cells of patients who had received a thalassaemia gene from one parent and the sickle-cell gene from the other contain more Hb S than Hb A—that is, the reverse of what is found in individuals with the sickle-cell trait, who have considerably more Hb A than Hb S in their red cells (see Figure 8).

This critically important observation suggested that the action of the thalassaemia gene was to reduce the amount of Hb A relative to Hb S, or, since the sickle-cell mutation was now known to affect the β globin chain, to reduce the amount of β^A relative to β^S chains. However, at about the same time research about further families was published that showed no interaction of this type—that is, some of those who were doubly affected with the gene for Hb S and thalassaemia showed the same amount of the variant haemoglobin as those who were simply heterozygous for Hb S. Equally important, a series of family studies begun at this time and extended into the early 1960s showed that, if some individuals with sickle-cell thalassaemia were married to a normal person, their offspring inherited either the sickle-cell trait or the thalassaemia trait; they were never normal or doubly affected, like their doubly affected parent. This suggested that the sickle-cell gene and the particular thalassaemia gene are alleles—that is, they both occur at the same genetic locus on the same pair of parental chromosomes. Again, since the sickle-cell mutation involves the β globin gene, this was clear evidence that there is at least one form of thalassaemia, β thalassaemia, that results from defective β globin chain production. In other families, however, the sickle-cell gene and the thalassaemia gene seemed to be passed on quite independently, and any combination of the genes could be found in the offspring of doubly affected people. Could this form of thalassaemia that did not interact with Hb S as just described be due to a mutation of the α globin genes?

Hence, by the late 1950s there was a series of apparently disparate observations of the haemoglobin patterns of patients with thalassaemia or of those who had inherited both thalassaemia and a structural haemoglobin variant like Hb S. Many severely affected children had large amounts of fetal haemoglobin that persisted after the first year of life, and in some but not all cases their parents had a raised level of Hb A_2. Studies of patients who had inherited both thalassaemia and Hb S, sickle-cell thalassaemia, showed two distinct patterns of haemoglobin; some had a reduced level of Hb A compared with Hb S, while others showed similar levels to those found in Hb S carriers. Patients with Hb H thalassaemia had variable levels of the β_4 molecule, Hb H, in their blood. Similarly, babies born with Hb Bart's had thalassaemic blood pictures. These findings, and those in patients with sickle-cell thalassaemia and their families, suggested that there must be more than one form of thalassaemia, at least one affecting the production of β globin chains, the other (possibly) affecting the output of α chains.

Early Hypotheses Regarding the Basis for Defective Globin Chain Production

As soon as it started to become apparent that thalassaemia might reflect a disease of adult haemoglobin production, thoughts started to turn to the possible mechanisms involved. In 1953 Itano discussed the significance of the finding that the level of Hb S in individuals with the sickle-cell trait—that is, carriers for the Hb S gene—have less Hb S than Hb A. He had observed that the level of Hb S in carriers falls into several modes that appeared to be genetically determined, a finding that led him to propose his 'structure/rate' hypothesis: that the rate of synthesis of a particular protein is in some way related to its primary structure.

In 1954 Pauling developed and extended this concept, suggesting that thalassaemia might result from the production of an abnormal haemoglobin with properties so close to those of normal adult haemoglobin that subtle differences might have escaped detection. He also suggested that, because the thalassaemia allele interferes with the manufacture of normal haemoglobin, but does not affect the manufacture of abnormal haemoglobin, as observed in some patients with sickle-cell thalassaemia, it must occupy the same locus on the chromosome as the alleles for other abnormal haemoglobins. He went on to speculate that the 'thalassaemia gene' might be responsible for the production of an abnormal haemoglobin of such a nature as to prevent the inclusion of haem, the ring structure that carries oxygen.

In 1957 Itano extended and refined the structure/rate hypothesis and wrote 'thalassemia mutants at the hemoglobin locus are analogous to the mutants for abnormal hemoglobins, differing in their failure to alter the net charge of adult hemoglobin and in the greater inhibition they exert on the net rate of synthesis'. Although these hypotheses were not to stand the test of time, they were valuable in that they focused the attention of many workers on what might be the underlying defects in haemoglobin synthesis in thalassaemia, a topic of broad debate for the next fifteen years until it was possible to analyse these diseases at the molecular level.

A Synthesis: The Concept of α and β Thalassaemia

One of the main spurs to the production of a useful working model of the general nature of the thalassaemias was the publication of both the papers and the discussions that followed from several major international meetings that were held in the late 1950s. In September 1957 a Symposium entitled 'Abnormal Haemoglobins', organized by the Council for International Organizations of Medical Sciences under the auspices of UNESCO and the World Health Organization, was held in Istanbul. Many of the observations described in the previous sections were discussed, but perhaps what is most remarkable about this meeting was the descriptions of abnormal haemoglobins and thalassaemia by authors from outside the USA and Europe, including Turkey, Iran, several parts of Africa, India, Ceylon (as it was then called), Thailand, and Indonesia. Of particular interest was the fact that the different interactions between the abnormal haemoglobins and thalassaemia appeared to be widely distributed in Africa and parts of the Middle East, while in South East Asia it was already becoming clear that the type of thalassaemias associated with the production of Hb H is particularly common.

In May 1959 the CIBA Foundation held a symposium, 'Human Biochemical Genetics in Relationship to the Problem of Gene Action', in Naples, which was attended by most of the key workers in the haemoglobin field of the time. At this meeting the structure/rate hypothesis and the genetic heterogeneity of thalassaemia were aired in great detail. From the lengthy published discussions after the papers it is clear that many workers were now beginning to think of the thalassaemias in terms of a heterogeneous group of disorders that might involve the production of the α or β chains of haemoglobin. For example, the Italian geneticist Ceppellini, discussing the genetics of thalassaemia, wrote: 'thus two genetically independent varieties of thalassaemia, α type and β type, could be visualized.' At the same meeting the extremely limited evidence that existed at the time for allelism between β thalassaemia and the β globin genes was also reviewed.

In 1959 Ingram and Stretton reviewed and extended these ideas in a paper in *Nature* that proposed a model for the genetic basis of the thalassaemias. They suggested that there were two major classes, α and β, in the same way as there are two major types of structural haemoglobin variants. They based this notion on published pedigrees and explained quite elegantly the interaction between β thalassaemia and β chain haemoglobin variants, and α thalassaemia and α chain variants. They went on to interpret the synthesis of Hb H as reflecting an inherited defect in α-chain synthesis, resulting in excess β chain production and hence β_4 molecules.

While agreeing with Pauling and Itano that the reduced rate of α or β chain synthesis might be due to a 'silent', undetectable, mutation in the haemoglobin genes, Ingram and Stretton also proposed an alternative explanation, the 'tap' hypothesis. This suggested that the defect might lie not in the structural gene but in the area of DNA in a connecting unit preceding it. Presumably this inferred that the connecting units might have a regulatory role in the production of globin chains.

In 1961 Itano and Pauling wrote a letter to *Nature* in which they suggested that the genetics and chemistry of haemoglobin that had been reported since 1957 had not substantially altered their inferences about the general nature of thalassaemia. Rather unfairly, they added that Ingram's papers were remarkable for the extent to which the custom of giving pertinent reference to the ideas of previous workers had been ignored! On rereading Ingram and Stretton's seminal paper, it is difficult to substantiate this criticism. Their paper was published in *Nature* in December 1959. Undoubtedly many of the conclusions and the hypotheses in their paper had been widely discussed by many workers, notably at the Naples Symposium in May of the same year. Granted, the concepts that were summarized in the *Nature* paper were the result of the ideas of many different workers from diverse fields coming together at an appropriate time. But Ingram and Stretton freely acknowledged their contributions at the end of their *Nature* paper, and their extensive list of references excludes none of the key discoveries that led to the development of the concept of α and β thalassaemia.

The End of the Beginning

The period between 1950 and 1960 was a time of extraordinarily imaginative and interdisciplinary thinking in a field that was moving extremely rapidly. The basic concept of there being two sorts of thalassaemia, the α and β thalassaemias, formed the linchpin for future investigation of the disease. Already information was coming from many sources that indicated that the thalassaemias are disorders of extremely widespread occurrence, ranging from Africa, through the Mediterranean, the Middle East, the Indian subcontinent to South East and East Asia. The title of a paper published in 1954, 'Mediterranean Anemia: A Study of 32 cases in Thailand', and its authorship, as well as offering a flavour of the reports that were appearing from all over the world at the time, also emphasize the remarkable speed at which international collaborations were evolving and the rapid pace of development in the field. The first author of this paper, Virginia Minnich, was a talented laboratory technician who worked with Carl Moore in St Louis. Soon after the end of the Second World War, the haematology group at St Louis began to form close connections with Thailand, and Minnich spent some time there helping to establish a strong haematology laboratory; the discovery of thalassaemia in the Thai population followed. This type of collaboration was to evolve in many parts of the world and was undoubtedly one of the major reasons for the rapid development of knowledge about thalassaemia from the 1950s onward. But there were several other developments of equal importance.

As we saw earlier in this chapter, the specialist meetings in the haemoglobin field that were held during the 1950s were an absolutely critical catalyst towards the exchange of new information and ideas that were to lay the basis for the future development of the thalassaemia field. Except for the French Society of Haematology, which was founded in 1931, there were no other specialist societies of this type formed until after the

Second World War. The International Society of Haematology was founded in 1946 in Dallas, Texas, and in the following year the European Society of Haematology was founded in Pavia. In 1957, and after lengthy discussions in Boston the previous year, the American Society of Hematology was founded, and several other national societies were formed at about the same time. Thus by the late 1950s there were many opportunities for the international exchange of research in the thalassaemia field in particular and in haematology in general.

These developments were accompanied by the appearance of increasing numbers of English-language journals in haematology and related fields. It was this trend, probably more than any other, that broke down the communication barrier between research workers in southern and northern Europe and the USA that had lasted since the beginnings of the thalassaemia field. For example, the journal *Acta Haematologica* was first published in 1948 and carried an extensive review of Italian work in the thalassaemia field by Marmont and Bianchi in its first volume. An equally important review was published by Chini and Valeri in the fourth volume of the American journal *Blood*, which was first published in 1946. By this time there were also several journals in the field of genetics; the first account in an English-language journal of the extensive genetic studies that had been carried out by Silvestroni and Bianco and their colleagues in the 1940s and 1950s appeared in a paper by Bianco and colleagues in *Annals of Eugenics* in 1952. Until these papers appeared, the extensive Italian research in the field had been published in Italian in local journals that were not widely available. Many seminal papers from workers in Greece and other European countries also started to appear in English-speaking journals at about the same time.

There is no doubt that this extremely rapid increase in international communication in the 1950s was a major factor in the rapid development of haematology and in the thalassaemia field over this period. Curiously, however, some discrepancies still remain to this day. In the plethora of textbooks of haematology that appeared in the latter half of the twentieth century and later, Thomas Cooley is always cited as the first to recognize thalassaemia; the fact that the disease was also identified independently by workers in Italy at about the same time is never mentioned. The reason for this omission is not clear, although, at least in part, it may reflect the parochialism that crept into a lot of scientific writing during its years of rapid expansion.

IV

THE DIVERSITY AND PATHOLOGY OF THE THALASSAEMIAS

Following the discovery of the structure of DNA by Watson and Crick in 1953, the new science of molecular biology began to expand at an extraordinary rate. By the early 1960s the genetic code was being broken, and, with the discovery of messenger RNA and the development of *in vitro* approaches for studying protein synthesis, a picture of how proteins are made and assembled was starting to emerge. As early as the 1960s thoughts were already turning to how knowledge from this exciting new field might start to be applied to the study of the thalassaemias. But a great deal more groundwork would be required before this became possible.

During the 1960s there was a consolidation of knowledge about the genetic control and diversity of the thalassaemias and a gradual understanding of how a defect in the synthesis of one or other of the globin chains might lead to the protean manifestations of a disease that was now being seen increasingly in countries all over the world.

The β Thalassaemias

By the early 1960s sufficient numbers of homozygotes and heterozygotes with β thalassaemia had been studied to establish a clear picture of the haemoglobin constitution in this disorder. Although the findings in homozygotes were not always easy to interpret, because the patients were receiving blood transfusion, it appeared that the fetal haemoglobin level could range widely between 10 and over 90 per cent of the total haemoglobin; the Hb A_2 values were inconsistent but usually normal and certainly not raised to the level found in heterozygotes. In large series of heterozygotes the HB A_2 level ranged over 3.5–6.5 per cent, although, as in many of the earlier studies, there appeared to be a few carriers with values in the normal range.

A key issue at this time was whether it would be possible to reproduce the findings in the few studies reported earlier that indicated that the genetic determinant for β thalassaemia is an allele of the gene for the β globin chains. As before, this depended on studies of the children of matings between compound heterozygotes for β thalassaemia and β globin variants such as Hb S or Hb C. By 1964 it was possible to analyse many offspring of this type, and it was confirmed that they were carriers of either β thalassaemia or Hb S; none was normal and none was doubly affected like his or her parent. There was now no doubt that the β thalassaemia and β globin genes—that is, the genes that carry the sickle-cell or Hb C mutations—are alleles.

In the thalassaemia field, as in many branches of science, important advances followed the development of new technology at least as commonly as new ideas. In the early 1960s improved methods for separating and analysing haemoglobin variants became available. For example, using a two-dimensional paper-agar gel electrophoretic system, it was possible to demonstrate the complete absence of Hb A in some cases of homozygous β thalassaemia. Starch gel electrophoresis, developed by Oliver Smithies for the separation of serum proteins in 1959, was adapted for haemoglobin analysis at the beginning of the 1960s. This technique offered a major advantage over filter-paper electrophoresis, particularly for detecting minor haemoglobin variants or fractions. When this approach was applied to the study of patients with sickle-cell or Hb C thalassaemia, it was found that they fall into two groups: those who produce Hb A, usually in the 20–30 per cent range, and those in whom no Hb A can be detected. Furthermore, the former group appeared to show further heterogeneity; in a few cases, very much lower levels of Hb A were found. In addition, the ability to produce Hb A always seemed to run true within families.

It appeared, therefore, that there must be at least two distinct forms of β thalassaemia, in which there is either a reduced level of β chain production or no β chain production. These conditions were given the names β^{o} and β^{+} thalassaemia, respectively; the term β^{++} thalassaemia also came into limited use to describe β thalassaemia with relatively high levels of Hb A.

The Discovery of Conditions Related to β Thalassaemia

Throughout the 1960s, studies of the haemoglobin constitution of parents of children with severe forms of thalassaemia continued to provide information about the heterogeneity of this disease. In 1961 there were reports of β thalassaemia carriers who had normal levels of Hb A_2 and unusually high levels of Hb F, in the 15 per cent range. Hitherto, β thalassaemia carriers had been identified by their raised levels of Hb A_2 and only slightly raised levels of Hb F, usually only about twice the normal level of Hb F in adults, or about 0.5–3 per cent. Since this new variant of thalassaemia interacted with typical β thalassaemia associated with a high Hb A_2 to produce a more serious clinical disorder in compound heterozygotes—that is, children who had inherited the high Hb A_2 β thalassaemia gene from one parent and a gene for a high level of fetal haemoglobin from the other—it appeared to be a form of β thalassaemia. Initially it was called the 'high fetal haemoglobin variety of β thalassaemia', or 'F thalassaemia'. The early descriptions of this condition came mainly from the Mediterranean region and Africa, but it soon became apparent that it also occurs in Asia. By the mid-1960s it had collected further synonyms, including 'normal HbA_2 β thalassaemia' and 'β thalassaemia type 2'.

Between 1968 and 1970 several individuals homozygous for F thalassaemia were discovered. In each case they had the clinical picture of a mild form of β thalassaemia and had 100 per cent fetal haemoglobin, with no Hbs A or A_2. Hence this condition could now be defined as a variant of β thalassaemia in which there is an absence of β *and* δ chain production; therefore, it was renamed yet again, this time as 'δβ thalassaemia'.

In the early 1960s there was another twist involving the δβ thalassaemia story. In 1961 the Boston geneticist Park Gerald discovered a family with a thalassaemia-like disorder in which one parent and four relatives of a child with the clinical picture of severe thalassaemia, and the affected child, carried about 10 per cent of an abnormal haemoglobin that behaved on electrophoresis in a similar way to Hb S. In 1962 Corrado Baglioni, a student of Vernon Ingram, carried out an extremely elegant analysis of the structure of this new haemoglobin variant, now called Hb Lepore after the family name of the child in whom it was first reported. He found that it has normal α chains combined with non-α chains that consist of portions of both δ and β chains. He suggested that, since the β and δ chain genes are likely to be linked—that is, adjacent on the same chromosome—a prediction that was later found to be true, there must have been slippage and unequal crossing over between the β and δ genes during meiosis—when parental strands of DNA become opposed in germ cells. The outcome of a genetic accident of this kind was the production of the δβ fusion genes for the δβ chains of Hb Lepore. Hence, Hb Lepore thalassaemia might be a form of δβ thalassaemia in which both δ and β globin chain synthesis is absent, just like the more common form of δβ thalassaemia described above. In fact, this prediction was proved to be correct when patients who are homozygotes for this condition were discovered later and found to make no Hb A but only Hbs F and Lepore.

Although Hb Lepore thalassaemia was subsequently found in many other populations, it is a rare form of thalassaemia. This does not undermine the importance of Baglioni's work, however; close on twenty years before it was possible to analyse the thalassaemias at the DNA level, it had been possible to describe the molecular basis for one form of the disease.

Yet another condition related to δβ thalassaemia was characterized at about this time, although its discovery came from a completely different source than the other forms of β and δβ thalassaemia. In 1955 Edington and Lehmann described a Nigerian patient with an unusually mild form of sickle-cell anaemia characterized by particularly high levels of fetal haemoglobin. A family study disclosed that this condition appeared to result from the inheritance of the sickle-cell gene from one parent and a gene associated with the production of approximately 25 per cent fetal haemoglobin in adult life from the other. Similar results were reported from Uganda and from the analysis of patients with mild forms of sickle-cell anaemia in the United States. For the want of anything better, this condition was named 'hereditary persistence of fetal haemoglobin', or HPFH.

In 1963 an extensive report of families of this type by Conley and his colleagues at Johns Hopkins Hospital, Baltimore, confirmed that the heterozygous state for HPFH is characterized by fetal haemoglobin levels in the 15–25 per cent range and that there are no haematological abnormalities. Those who inherited the HPFH gene together with Hbs S or C produced no Hb A, but again, the levels of Hb F were unusually high. Even more importantly, the offspring of matings between such persons and normal people were heterozygous for either HPFH or the β globin variant, suggesting allelism or close linkage between the β globin gene and that for HPFH. A few years later several individuals homozygous for HPFH were discovered, all of whom had 100 per cent fetal haemoglobin with no Hbs A or A_2. Although they were clinically normal, their red cells were both small and poorly haemoglobinized. Thus it appeared that HPFH is an extremely mild form of δβ thalassaemia in which defective β chain production is almost, but not entirely, compensated for by increased Hb F synthesis.

If all this was not enough, evidence appeared during the 1960s for even further heterogeneity of the β thalassaemias. Forms of otherwise typical β thalassaemia were identified in which carriers had unusually high levels of Hb F or Hb A_2, and several families were identified in which a moderately severe form of a β thalassaemia-like condition was inherited in a dominant fashion—that is, the clinical disorder was due to a single defective β globin gene.

So, in the ten years that followed the classification of the thalassaemias into α and β forms, substantial evidence for the heterogeneity of the β thalassaemias was obtained. It was now apparent that there must be many different types of mutation that can affect the β globin genes themselves, and another and equally diverse set of lesions that can underlie the defective synthesis of *both* δ and β chains.

The α Thalassaemias

The exploration of the α thalassaemias between 1960 and their first analyses by the tools of molecular biology in the mid-1970s is a good example of the confusion that can arise in a research field if one or more of the critical premises on which it is based are uncertain or even wrong. In retrospect, much of the confusion that arose during this period simply reflected lack of knowledge about the precise way in which the production of the α globin chains is genetically determined. It was not until 1970 that a critical family was reported from Budapest in which two α globin chain haemoglobin variants segregated independently, proving that in human beings the α globin genes are duplicated—that is, there are a pair, αα, on each of the homologous chromosomes. Without this knowledge it was extremely difficult to interpret some of the findings in patients that were being reported from all over the world with different forms of α thalassaemia.

Knowledge of the α thalassaemias accumulated from a variety of different directions and clinical observations, as had been done for the β thalassaemias. One important source of information was the study of Hb Bart's in the blood of newborns. Since Hb Bart's has the structure γ_4, and has no α chains, it was already thought that its presence must represent some form of defective α chain production. In 1960 Lie-Injo and Jo described a stillborn Indonesian infant whose haemoglobin consisted mainly of Hb Bart's. Further cases were reported from Malaya, and the condition became widely recognized in South East Asia over subsequent years. It appeared that these babies had inherited an extremely severe form of α thalassaemia, which caused them to produce very high levels of Hb Bart's and which led to stillbirth associated with severe intra-uterine anaemia and hypoxia, and hence the clinical picture of hydrops fetalis. Most of these babies died in the last third of pregnancy or, if live born, took a few gasps and then rapidly expired. The picture of fetal hydrops was, of course, well known in other conditions of profound intra-uterine anaemia, including Rhesus blood group incompatibility between mother and fetus. It is characterized by gross pallor and marked swelling of the abdomen and limbs because of fluid retention, presumably reflecting increased permeability of the blood capillaries consequent on profound oxygen shortage owing to profound anaemia.

While these observations were being made in Asia, reports from Africa and the USA suggested that there was a high frequency (in the 5 per cent range) of newborn babies of African origin with increased levels of Hb Bart's. A curious feature, however, was that the variant usually disappeared over the first six months of life and was not replaced by the adult counterpart of Hb Bart's—that is, Hb H, or β_4. These babies also had small, poorly haemoglobinized red cells, which persisted during later development in many cases. Perhaps, it was argued, this moderate increase in Hb Bart's at birth might represent a milder form of α thalassaemia. A hypothesis was developed along the lines that the neonatal period, at the time of the switch from fetal to adult haemoglobin production and when both γ and β chains are competing for available α chains, may be a particularly valuable period to recognize a mild deficiency of α chains, which is reflected by a small excess of γ chains. Once the switch from γ to β chain production is complete, the deficiency of α chains might be too small to lead to the production of detectable amounts of Hb H. Perhaps, it was argued, there is an extremely mild form of α thalassaemia in African populations that is not severe enough to produce the picture of Hb Bart's hydrops.

Further evidence in support of this idea came from the discovery of an Afro-American woman who was a carrier for an α chain structural haemoglobin variant called Hb I. Surprisingly, this patient's red cells contained approximately 70 per cent Hb I; as is the case for carriers of β structural haemoglobin variants like HbS, carriers for α chain variants usually have more Hb A than the variant. It seemed possible, therefore, that this woman, in addition to the structural variant, had inherited a mild α thalassaemia gene, a supposition that was supported because she had haematological changes consistent with thalassaemia. Her children all had small amounts of Hb Bart's at birth, further evidence for the existence of a mild form of thalassaemia among those of African descent.

During this period reports of the moderately severe form of thalassaemia associated with the production of varying amounts of Hb H, or β_4, which was also thought to be a form of α thalassaemia, started to appear from all over South East Asia and to a lesser degree from the Mediterranean region. The genetic transmission of this condition was extremely complex and varied considerably from case to case; there were even families reported in which it appeared to be passed on directly from one generation to another. However, by the mid-1960s it appeared that it might result from the interaction of a severe α thalassaemia gene with a second gene that was not detectable by techniques that were available at the time. It was reasoned that this 'silent' gene might also be an extremely mild α thalassaemia gene—further evidence suggesting genetic heterogeneity of the α thalassaemias.

In short (and for readers who have struggled this far with this complicated story), by the mid-1960s it was clear that there must be a very severe α thalassaemia gene that, since by then it was known that the same α globin genes control the production of α chains in both fetal and adult life, would lead to stillbirth in the homozygous state. The moderately severe form of α thalassaemia, Hb H disease, which was compatible with survival to adult life, appeared to result from the inheritance of at least one severe α thalassaemia gene and possibly a second that is completely silent in carriers. Finally, it appeared that there was a mild form of α thalassaemia that occurs at high frequencies in African and Oriental populations.

By now the scene was set for workers in a population with a very high incidence of the different clinical types of α thalassaemia to try to put all these apparently disconnected facts together. These critical studies came from a group in Bangkok led by Prawase Wasi. This team played a major role in furthering our understanding of thalassaemia, both in Thailand and elsewhere in South East Asia.

The Thai workers took two different though related approaches to the problem of α thalassaemia. At first, they made extensive studies of the level of Hb Bart's in Thai newborns to see if they could demonstrate any segregation of values of levels of the variant at birth. They found that the relative amounts of Hb Bart's could be graded into traces, small amounts, and moderate amounts, corresponding to 1–2, 5–6, and 25 per cent of Hb Bart's, respectively. Assuming that these different levels reflected different degrees of severity of α chain deficiency, they used the terms α thalassaemia 1 for the more severe α thalassaemia allele and α thalassaemia 2 for the milder one, and suggested that the concentration of Hb Bart's in cord blood of 100, 25, 5 and 1–2 per cent represent α thalassaemia-1 homozygosity, α thalassaemia 1/α thalassaemia 2 (Hb H disease), α thalassaemia-1 trait, and α thalassaemia-2 trait, respectively.

Using this as a working hypothesis, the Thai group then proceeded to investigate the frequencies of different α thalassaemia genes in northern Thailand. For this purpose they applied the Hardy–Weinberg law, a method for predicting genotype frequencies on the basis of gene frequencies under the assumption of random mating in the absence of selection. They found that the distribution of the frequencies that they obtained were compatible with the genes for α thalassaemia 1 and α thalassaemia 2 being alleles or closely linked. Furthermore, their findings provided strong evidence that Hb H disease results from the inheritance of both these genes. Further evidence that this is the case was obtained when they examined thirty-one offspring of patients with Hb H disease and found that all of them had Hb Bart's; there was a tendency for the levels of Hb Bart's to fall into two groups with values of 1–2 and 5–6 per cent, respectively. As mentioned earlier, one of the difficulties in understanding the genetic transmission of Hb H disease had been the many reported examples of 'parent-to-child' transmission. If the two α thalassaemia genes are alleles, this must result from a mating between a patient with Hb H disease and a carrier of either α thalassaemia 1 or 2. Wasi and his colleagues estimated that, in order to account for the observed incidence of parent-to-child transmission in Thailand, the overall incidence of the two α thalassaemia genes must be about 21 per cent; in the cord-blood study cited above, there was an overall incidence for α thalassaemia 1 and 2 of 20.4 per cent, a remarkably close agreement to the predicted figure.

Shortly after these seminal studies from Thailand had been reported and as mentioned earlier, a Budapest family with two α chain variants was reported. It was now clear that there must be two α globin chain loci on each pair of homologous chromosomes in humans. Hence it was now possible to explain the basis for the two variants of α thalassaemia defined in the Thai studies based on lesions of one or other of these loci.

Although there had been major difficulties in interpreting the different threads of the α thalassaemia story, by 1970 it was clear that there were two common forms of α thalassaemia in many Asian populations and a milder form of the condition that occurs at very high frequencies in Africa and elsewhere. The scene was now set for investigating these conditions at the molecular level, as the appropriate techniques became available in the 1970s.

It was now apparent, therefore, that there must be many different forms of thalassaemia (Table 2). However, they all

TABLE 2 **The different varieties of thalassaemia**

β thalassaemia	Defective or absent β chain synthesis. Sometimes subdivided into β^{o}, β^{+} or β^{++} thalassaemia depending on the degree of reduction in β chains. Homozygotes or compound heterozygotes who have inherited a different mutation from each parent usually have severe anaemia. Heterozygotes have small red cells and raised levels of Hb A_2.
δβ thalassaemia	No δ and β chain synthesis. Homozygotes have 100% Hb F and moderate anaemia. Heterozygotes have small red cells and raised Hb F levels.
Hb Lepore thalassaemia	No δ or β chain synthesis. Produce small amounts of δβ fusion chains combined with α chains, Hb Lepore. Homozygotes have severe anaemia and Hbs F and Lepore. Heterozygotes have small red cells and 10–20% Hb Lepore.
α^{o} thalassaemia (or α thalassaemia 1)	No α chain synthesis. Homozygotes die *in utero* or at birth. Heterozygotes have small red cells and normal Hb A_2 levels. Coinheritance with α^{+} thalassaemia causes Hb H disease.
α^{+} thalassaemia (or α thalassaemia 2)	Mild defect in α chain synthesis. Homozygotes have small red cells with normal Hb A_2 levels. Heterozygotes have no haematological changes.
Hereditary persistence of fetal haemoglobin (HPFH)	Many different types. Those homozygous for the variety owing to deletions of the β and δ globin genes have 100% Hb F and small red cells.

(*Cont.*)

TABLE 2 (Continued)

	Heterozygotes have normal red cells and 15–25% F. The other varieties show no haematological changes.
Dominant β thalassaemia	Moderate anaemia and abnormal red cell morphology owing to a single defective β globin gene.

seemed to share the same kind of clinical and haematological abnormalities as were originally described by Thomas Cooley: the varying degree of anaemia, small poorly haemoglobinized (hypochromic) red cells, which varied widely in their shape, and enlargement of the spleen. Severe forms were also associated with the skeletal deformities described by Cooley. Quite remarkably, however, there was wide variation in the degree of anaemia and associated clinical findings, even in the same type of thalassaemia. Although by the end of the 1960s it had also become apparent that these clinical pictures could result from many different forms of α or β thalassaemia, what was still not clear was how defective α or β chain production, if indeed this was the basis for these conditions, could bring about these remarkably diverse patterns of disease.

The Relationship between Defective Haemoglobin Synthesis and the Clinical Picture of Thalassaemia

By the early 1960s, although there were the beginnings of an understanding of the genetic basis of the thalassaemias, it was still not clear how a single defective gene could lead to the extraordinarily diverse clinical features of the disease. It was now necessary to understand the basis not only for the profound anaemia of severe cases but also for the other features first described by Thomas Cooley, including deformities of the skull and face, massive enlargement of the spleen and liver, and pigmentation of the skin, to mention but a few. After all, there are many forms of severe anaemia but they are not associated with this striking constellation of clinical manifestations. Words such as 'pleotrophic' were bandied about to explain the multiple effects of the genes for thalassaemia, concepts that led nowhere, even if they were of ephemeral comfort to those who proposed them.

Early thoughts on the nature of the underlying pathology of thalassaemia that focused on the haematological manifestations were also clouded by the extraordinary diversity of the changes that were described. While most of the early workers in the field stressed the importance of the haemolytic component of the anaemia—that is, a shortened red-cell survival—as early as 1932 Cooley and Lee felt that this was not enough to account for the profound anaemia that they saw in their children. Recognizing that the bone marrow appearances were also strikingly abnormal, they suggested that the anaemia may be the result of a metabolic disturbance 'which compels the tissues to make bricks without straw'. In 1949 Perosa listed the possible causes of the anaemia as impaired union of iron with porphyrin, defective porphyrin synthesis, or a lack of or abnormal synthesis of globin. In a review written about the same time by the Italians Marmont and Bianchi, the diverse haematological and related aspects of the disease were again stressed, including increased iron loading of the tissues, faulty red-cell production in the bone marrow, high iron levels in the blood associated with pale and deformed red cells that looked as though they were deficient in iron, and a shortened red-cell survival. They too found it impossible to ascribe these diverse findings to a single mechanism. The best they could do was to describe the disorder as a form of metabolic dyserythropoiesis, a term first used by the British haematologist Janet Vaughan in 1948.

Some genuine progress towards an understanding of the pathophysiology of thalassaemia was made in the mid-1950s when it became possible to study red-cell production—the turnover of red-cell precursors in the bone marrow and their survival in the blood—using radioactive isotopes. Sturgeon and Finch were the first to carry out careful studies of this type on patients with thalassaemia. Their work showed that there was a marked degree of ineffective erythropoiesis in this disorder—that is, there was a major proliferation of the precursors of red blood cells in the marrow but many of them were destroyed before they even reached the blood. Indeed, further studies showed that the degree of ineffective red-cell production in thalassaemia is probably greater than in any other disease. Taken together with the massive expansion of the bone marrow, this work suggested that there is a major drive to overcome the anaemia but that the expanded marrow is incapable of effective red-cell production. But until the early 1960s there was no clue as to how a defect in haemoglobin synthesis could produce this remarkable picture, or the shortened lifespan of the relatively few red cells that did reach the peripheral blood.

The Discovery of Inclusions in the Red-Cell Precursors

In 1963 the Greek haematologist Phaedon Fessas, whom we have met in discussing early developments in the thalassaemia

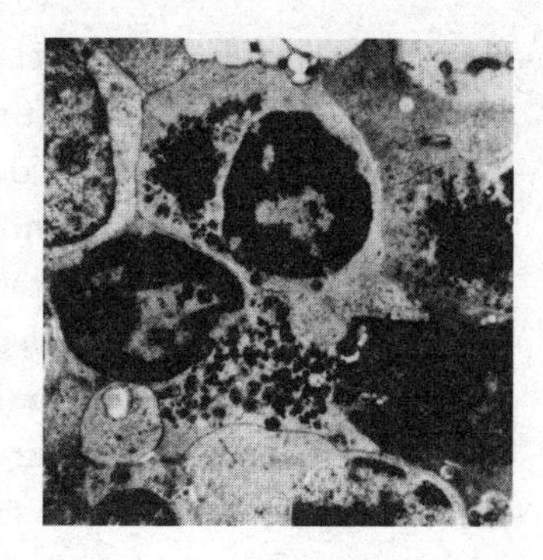

10. Ragged inclusion bodies in the red-cell precursors in the bone marrow of a patient with β thalassaemia. The small inclusions are seen as well as the large dark staining objects that are the cell nuclei. (*From an electron micrograph kindly prepared by Dr Sunitha Wickramasinghe.*)

field, made the extremely important observation that there are large, ragged inclusion bodies in the red-cell precursors of patients with β thalassaemia (Figure 10). They are also present in the peripheral blood of these patients, but only in those who had had their spleen removed. Since the staining properties of these inclusion bodies suggested that they contain haemoglobin in one form or another, Fessas suggested that they might be α chains; he reasoned that in β thalassaemia there might be an excess production of α chains that might precipitate and produce these bodies. However, because as yet there was no evidence that there was imbalanced globin production in thalassaemia, he also suggested that they might be aggregates of abnormal β chains or complete but unstable haemoglobin molecules.

Readers might wonder why it took close on forty years from the first description of thalassaemia before these inclusion bodies were identified, particularly since all Fessas required to make this seminal observation was a microscope and a few simple stains! However, a reading of earlier papers that dealt with the morphology of the marrow and blood in this disease shows very little evidence that they were observed before 1963. Granted, in the mid-1950s there were reports of single inclusion bodies in the red cells, though not in the bone marrow of patients with Hb H thalassaemia who had undergone splenectomy. It was suggested at the time that Hb H is intrinsically unstable and precipitates during the lifespan of the red blood cell, and that the shortened red-cell survival in this condition is due to the trapping of cells containing these inclusions in the spleen. Clearly, a similar mechanism might occur in β thalassaemia, leading to the shortened survival of such cells that are produced, a reasonable explanation for why Fessas observed inclusion bodies only in the blood of patients who had undergone splenectomy.

But what were these inclusions and what was their relationship to abnormal haemoglobin production in thalassaemia? The answers were to come only two years after Fessas's observations.

Haemoglobin Synthesis in Thalassaemia

In the early 1960s it was realized that to understand the pathophysiology of thalassaemia, how abnormal gene action at the globin loci could lead to the protean clinical manifestations of the disease, and to try to make at least some progress towards an understanding of the types of mutation that might underlie the different forms of thalassaemia, a more dynamic approach was needed. Therefore, thoughts turned to the possibility of trying to study haemoglobin synthesis in the red-cell precursors of thalassaemic patients using some of the *in vitro* experimental systems that were being developed at the time to study normal haemoglobin synthesis.

It had been known for several years that it is possible to incorporate radioactive amino acids into haemoglobin in red cells incubated *in vitro*, provided there are sufficient numbers of reticulocytes—that is, young red cells with the capacity to make haemoglobin—present in the blood samples. Much of this work had been carried out using the blood of rabbits that had been made anaemic with an agent that caused premature destruction of their red cells and hence that resulted in an extremely high reticulocyte count. Even in the early 1960s, a time when the ethics committees were much less demanding than they are today, it was not possible to use an experimental approach like this in humans. Clearly, experiments of this kind would be hindered by the relatively low reticulocyte counts in the blood of patients with thalassaemia.

Undaunted, several groups of workers in the late 1950s and early 1960s made use of these techniques to try to pinpoint the defect in haemoglobin synthesis in thalassaemia. The earliest studies were carried out with a view to analysing the synthesis of the haem fraction of the haemoglobin molecule in thalassaemic red cells. Although some rather non-specific abnormalities were found, it was difficult to equate these observations with the later findings of the haemoglobin patterns in patients with sickle-cell thalassaemia; a defect in haem synthesis would have affected both Hb A and Hb S equally and could not explain the reduced amount of Hb A in these doubly affected patients. Hence there was little further work on haem synthesis in thalassaemia in the 1960s, and thoughts turned to the development of approaches to study globin synthesis in these disorders.

By the early 1960s the basic principles of how proteins are synthesized had been determined, a topic to which we shall return in a later chapter. The information required to place the individual amino acids in a peptide chain—that is, a string of amino acids such as the α and β globin chains—is contained in a ribonucleic acid (RNA) structure called messenger RNA. The amino acids are transported to the latter, and the growing chain is held in position as it is synthesized by structures called ribosomes. In 1961, in a beautifully elegant experiment, Howard Dintzis demonstrated that globin chains are synthesiszd in an orderly manner, starting at the amino-terminal and proceeding to the carboxy-terminal end (in simple terms, from left to right!). The first experiments on globin synthesis in thalassaemia were carried out in New York by Paul Marks and his colleagues, who found that ribosomal function in thalassaemic cells was similar to that in normal reticulocytes. They also found that Hb F synthesis proceeds in a similar way in cells from thalassaemic and non-thalassaemic subjects but were not able to make further progress in defining the pattern of defective haemoglobin synthesis.

The central problem at this time was that there was no way of separating the α and β globin chains that could be used in experiments to determine their relative rates of synthesis. The methods that were available were extremely tedious, required large amounts of globin, usually resulted in poor yields of either the α or β chains, and therefore were totally unsuitable for analysis of the relatively small blood samples that were available from patients with thalassaemia. I was trying to solve this problem, and, using a variety of these inefficient methods, I was able to demonstrate unequal labelling of the α and β chains in thalassaemic reticulocytes, a finding that was also reported by Heywood and his colleagues in 1964.

The technique that enabled further progress to be made was developed by John Clegg and me in 1965. In fact, it evolved from work that Clegg had carried out on the separation of the chains of a completely different protein as part of his Ph.D. project. It was found that the peptide chains of globin could be fractionated quantitatively by chromatography on a specific cellulose in a strong buffered solution of urea. The key to the successful separation of the chains was the addition of 2-mercaptoethanol to the buffer, which prevents aggregation of the α and β chains by inhibiting bonding between them. Incidentally, the vile smell of this reagent did not enhance the popularity of our laboratory. However, by using this approach, if human reticulocytes are labelled *in vitro* and then lysed and the whole lysate converted to 'globin', it is possible to measure the total amounts of α and β chains synthesized, with the recovery of radioactivity and protein in excess of 95 per cent. As this technique was further improved, it also became possible to separate the γ chains of Hb F from the β chains of Hb A and hence to provide an accurate assessment of the overall relative rates of globin chain synthesis.

In the first description of the application of this method to the study of thalassaemia in 1965, it was possible to provide a reasonably complete picture of the pattern of normal globin synthesis and of abnormal globin synthesis in both α and β thalassaemia. It was found that in normal reticulocytes α and β chain production is almost synchronous, though there is a very slight excess of α chain synthesis. In β thalassaemia there is marked imbalance of globin chain production and, overall, a variable excess of α chains are produced (Figure 11). In Hb H disease, on the other hand, there is a marked deficit of α chain synthesis; as well as the β_4 molecules of Hb H, the red-cell precursors of patients with this disease contain a relatively large pool of free β chains, which are available to combine with

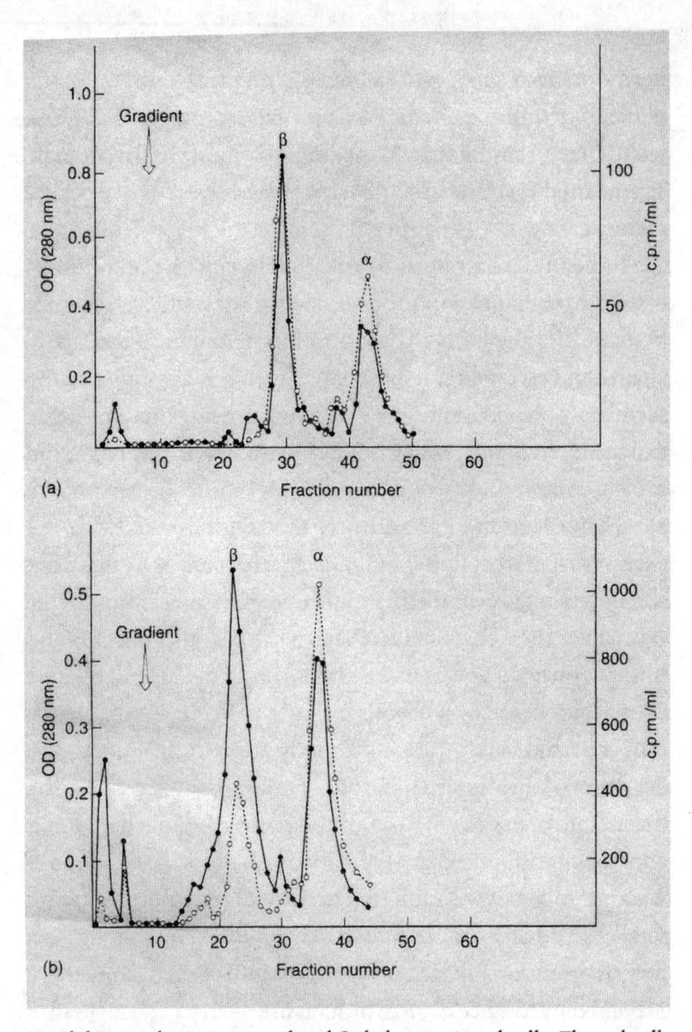

11. Globin synthesis in normal and β thalassaemic red cells. The red cells were incubated with radioactive amino acids, and then the globin chains were separated by chromatography. The continuous black line represents the protein fractions of the α and β chains, while the broken line represents the radioactivity incorporated into these chains during a period of *in vitro* incubation. The top graph shows normal haemoglobin synthesis, while the bottom graph shows the pattern of synthesis in homozygous β thalassaemia.

newly made α chains to produce Hb A. Similar results for patients with β thalassaemia major were obtained using this technique a year later by Bank and Marks, who also demonstrated mild globin chain imbalance in the red cells of β thalassaemia carriers.

The penny had dropped at last! Thalassaemia is not so much a disease of haemoglobin production as a disorder characterized by *imbalanced* globin chain production, with a marked excess of α chain synthesized in β thalassaemia and vice versa in α thalassaemia. In subsequent years this work was continued and expanded by several groups, notably Arthur Bank and Paul Marks in New York, and Sergio Pontremoli and his colleagues in Ferrara, as well as Clegg and myself. It was found, for example, that the excess α chains produced in β thalassaemia undergo two fates; some are rapidly degraded by the proteolytic enzymes of the red cell, while others become associated with the red-cell membrane. It was confirmed that in some forms of β thalassaemia there is a complete absence of β chain synthesis, while in others there is only a partial defect. It was also found that there is a marked degree of globin chain imbalance in other common forms of thalassaemia, such as sickle-cell and Hb E β thalassaemia. In another important outcome of this work, it was found that babies who are stillborn with the Hb Bart's hydrops syndrome have a total absence of α chain synthesis, which confirms that they are homozygous for a particularly severe α thalassaemia gene that prevents any form of α chain production.

Many groups attempted to dissect the patterns of globin chain production in carriers for the severe and mild forms of α thalassaemia. While there was clearly globin chain imbalance in these individuals, it was difficult to find clear-cut differences in the relative severity of the defect in α chain production between mild and severe carriers. In retrospect, this is almost certainly because of the technical difficulties involved in carrying out these experiments, characterized as they were with very low levels of isotope incorporation into the haemoglobin of these individuals because their red cells contained very few reticulocytes.

All these studies, while pinpointing the basic phenomena of imbalanced globin chain synthesis, still said nothing about the reasons for the defective production of α or β globin chains in thalassaemia. During the late 1960s Clegg and I carried out an extensive series of experiments based on the earlier studies of Howard Dintzis in which we attempted to measure the patterns of assembly of the α and β globin chains in both normal and thalassaemic red cells. The results suggested that there was no detectable abnormality in the assembly, initiation, or termination of globin chain synthesis in α or β thalassaemia, suggesting that, at least in cases in which some gene product—that is, globin chain—is synthesized, the defect might lie in a reduction in the amount of messenger RNA for the particular globin chain.

By the end of the 1960s, therefore, it had been clearly established that the thalassaemias are due to imbalanced globin chain synthesis and there were some hints, though no more than that, about the type of basic molecular defect that might be encountered when it became possible to analyse the α and β globin genes directly.

A Better Understanding of the Cellular Pathology

Once it was appreciated that the thalassaemias are diseases that are due to imbalanced globin chain production, it became possible to start to make some sense of the previously reported changes in the marrow and blood of patients with these diseases. Since in β thalassaemia there is a large excess of α chains produced that precipitate and form inclusion bodies in the red-cell precursors, the subsequent damage resulting from this process provided a mechanism for the remarkable degree of ineffective erythropoiesis and destruction of red-cell precursors in the bone marrow that had been observed ten years earlier. Similarly, the observation that in patients who had had their spleens removed these inclusions could be observed in the peripheral blood suggested that trapping of cells containing inclusion bodies in the spleen, or the removal of the inclusions in the spleen with damage to the red cells, might be at least one mechanism underlying the shortened red-cell survival that had also been observed in thalassaemia for many years.

At the same time as studies in haemoglobin synthesis were progressing, some interesting observations were made about the peripheral blood cells in β thalassaemia. In particular, it was noticed that the fetal haemoglobin that is produced in high levels in this disease is quite unevenly distributed between the red cells. How could this be? In 1966 Nathan and Gunn, building on the concept of imbalanced globin chain production, argued that, in β thalassaemia, those red-cell precursors that continue to produce γ chains will be relatively protected from the deleterious effects of α chain precipitation, because some of the excess α chains combine with γ chains to make Hb F. This mechanism would account for their earlier studies, based on *in vivo* labelling of red blood cells, that demonstrated that Hb F-containing cells survive relatively longer than Hb A-containing cells in the blood. In short, the ability to make Hb F in adult life appears to be restricted to certain groups of red-cell precursors, accounting for the heterogeneity of the distribution of Hb F in the red cells and also the longer survival of cells containing relatively high levels. They supported these suggestions with careful morphological studies of the red cells, finding that the better haemoglobinized cells with more Hb F had smaller inclusions. They also suggested that, in addition to causing abnormalities of cellular maturation, precipitating globin might interfere with red-cell membrane structure and function, leading to 'leakiness' of the red cells. This, together with the mechanical effects of inclusions on red cells in the circulation and spleen, might at least partly explain the haemolysis observed in the thalassaemias.

Although many of these ideas had to be refined and extended as more was learnt about the mechanisms of haemoglobin precipitation, research carried out over subsequent years confirmed that this general concept of the basis for the anaemia of thalassaemia is more or less correct. In short, the profound anaemia of severe thalassaemia is the result of globin chain imbalance, leading to a complex combination of ineffective erythropoiesis and shortened red-cell survival.

These concepts were also extended to explain some of the other remarkable clinical features of thalassaemia that had been observed since Cooley's first descriptions. Studies in the 1950s demonstrated that profound anaemia stimulates the production of the hormone erythropoietin, leading to the expansion of red-cell precursors in the bone marrow. In thalassaemia this leads to massive expansion of an ineffective erythroid marrow, resulting in severe bone deformities. Further work confirmed that the widespread iron deposition in the organs is the result of increased iron absorption, which occurs in the face of ineffective red-cell production in the bone marrow, together with the effects of blood transfusion. The pigmentation of the skin characteristic of thalassaemia could, at least in part, be explained by iron deposition. By 1970, therefore, it became unnecessary to ascribe mysterious pleotrophic—that is, unrelated—effects to the action of the thalassaemia genes; a basic defect in globin production could explain many of the widespread clinical manifestations and, incidentally, at least some of the remarkable clinical heterogeneity that had been observed increasingly over the years.

Further Thoughts about the Basic Defect in Globin Chain Production

At meetings and symposia held on abnormal haemoglobins in the early 1960s, notably those at Arden House in 1962 and Ibadan in 1963, the structure/rate hypothesis of the basic defect in globin chain production continued to be aired. At the Ibadan meeting, Itano recapitulated his original hypothesis and re-examined it against the background of the rapid acquisition of knowledge about the functions of DNA, RNA, and the genetic code. He explored the possibility that the varying affinity of putative abnormal thalassaemic β chains for α chains might play a role in the pathogenesis of at least some forms of β thalassaemia. He also suggested several alternatives, including rate-limiting steps in variant globin chain assembly. By now it was well established that protein synthesis depends on the appropriate amino acid being transported to the messenger RNA template by transfer RNAs, which are specific to each amino acid. Since the genetic code is degenerate—in other words, there is more than one code word for each particular amino acid in a peptide chain—it is possible that a base substitution could generate a messenger RNA codon that is identified by a transfer RNA that is in relatively low abundance. This, he argued, might lead to a reduced rate of globin production. He ended his review by concluding that the latter mechanism is the most likely explanation for defective globin chain synthesis.

Curiously, and by one of those quirks that makes the history of science so fascinating, there was no mention at the Ibadan symposium in 1963 that, at the meeting a year before at Arden House, Guidotti had announced that he had sequenced a considerable part of the β globin chain of a patient with β thalassaemia and found absolutely no abnormality in its amino-acid composition or sequence. Although as far as I know this work was never repeated or even published in detail, this observation, together with our studies on the assembly of the globin chains cited earlier and that were carried out at about the same time, finally put paid to the structure/rate hypothesis, which, by the late 1960s, no longer seemed tenable.

But there was no shortage of other speculations about the possible molecular basis for the thalassaemias. At a meeting held in New York in 1963 Ingram reviewed all the previous molecular models and added a few new ones for good measure. In a complex series of speculations he borrowed freely from the beautiful work that the French scientists François Jacob and Jaques Monod had carried out on the regulation of enzyme production in micro-organisms. It is not surprising that their work was freely adapted in the early 1960s by many other workers as a possible model for the regulation of globin-gene expression. Its sheer elegance suggested that, though a bacterium is a long evolutionary step from a human being, it is possible that similar mechanisms might be responsible for the regulation of protein synthesis in higher organisms.

In brief, what Jacob and Monod described was the operon. This consists of a group of linked genes of similar metabolic function, the activity of which is controlled by a closely linked gene called an operator. The action of the operator is always in *cis*—that is, on the same chromosome. The operator itself is under the control of what was originally called a 'regulator gene', the product of which is a specific repressor that acts both in *cis* and *trans*—that is, on both the same chromosome and on the opposite of a pair of homologous chromosomes. Because of its particular allosteric properties, the repressor can exist in several different states, depending on its interaction with various metabolites, thus allowing induction or repression of sets of genes of related function. The only thing that the human globin gene cluster has in common with an operon is that it consists of a group of linked genes of related function! But, given its wonderful versatility, it was possible to ascribe almost any observation in the haemoglobin field to a putative mutation involving either the operator itself, the regulator, or its repressor product. The only problem was that none of these speculations was open to any form of experimental verification at the time.

In 1963 the geneticist Walter Nance proposed a completely different mechanism for the cause of various forms of thalassaemia, based mainly on previous work by the geneticist Oliver Smithies on proteins called haptoglobins that bind haemoglobin. In short, Nance suggested that there might have been a series of unequal but homologous crossovers affecting the globin gene cluster. Readers will recall that this model is very similar to that described in an earlier chapter for the production of the Lepore haemoglobins.

By the end of the 1960s there was no shortage of models and hypotheses for the molecular basis of thalassaemia, but it was not obvious where to go next. Most of the earlier speculations seemed to be incompatible with the findings of haemoglobin synthesis that were made in the late 1960s, and, in the absence of any evidence in favour of the structure/rate hypothesis, it had finally been laid to rest. In effect, this was about as far as the protein field could take us in the study of thalassaemia; further progress required the new tools of the era of molecular biology, several of which became available at the beginning of the 1970s.

But before continuing the long saga of working out the molecular basis of the thalassaemias, we must take a brief respite and ask to what extent the progress that had been made towards an understanding of these diseases before 1970 had changed the lot of our patients.

V

EARLY IMPROVEMENTS IN THE MANAGEMENT OF CHILDREN WITH THALASSAEMIA

In the second edition of her monograph *The Anaemias*, written in 1936, Janet Vaughan described thalassaemia as a disease that is slowly and progressively fatal. There were no satisfactory forms of treatment; large doses of iron, liver, and liver extract were valueless. Blood transfusions were unavailing, and removal of the spleen was rarely followed by improvement. The majority of children died before the age of 9 years.

Except for a brief period of enthusiasm for the use of high doses of cobalt, this gloomy view of the management of thalassaemia, or the lack of it, did not change greatly for many years. However, with developments in blood-transfusion technology and the very slow evolution of a more rational approach to the use of splenectomy, there were gradual improvements in the symptomatic management of thalassaemia after the Second World War. At least some of these advances undoubtedly resulted from a better understanding of the pathogenesis of the disease, as outlined in previous chapters.

Blood Transfusion

The practice of blood transfusion has a long and often torrid history. Indeed, the first well-documented human transfusions took place in 1667, one in England and three in France. On 23 November of that year the Oxford physician Richard Lower, after several years of experiments on dogs, performed his first human transfusion before an audience at the Royal Society, London. His patient was a young clergyman, 'somewhat unbalanced whose brain was considered a little too warm'. After initial bleeding, he was connected via silver tubes to a sheep's carotid artery and received approximately 10 ounces of blood from the sheep. As described later in Pepys's *Diary*, he returned to the Royal Society six days later to tell the audience how much better he was feeling. In the same year Jean-Baptiste Denis in France carried out a similar series of transfusions, the third of which ended in the death of the patient. Later the patient's wife brought an action against Denis, claiming that the transfusion had killed her husband, a charge that was refuted when it emerged that the patient had been poisoned with arsenic by his wife!

Although many attempts at transfusion were made over the years, it was not until the discovery of blood groups by Carl Landsteiner in 1901, and the development of agents that would prevent blood clotting, that the modern era of blood transfusion began to evolve. Spurred on by the requirements of two world wars, this field developed rapidly over the first half of the twentieth century.

In the late 1940s and early 1950s reports began to appear about experience of the use of repeated transfusion in children with severe thalassaemia. As early as 1952 it had become clear at their autopsies that there was widespread damage to their organs because of the excessive amounts of iron that they had accumulated as the results of both transfusion and increased absorption of iron from their bowels.

The rather chaotic and totally unsatisfactory state of the management of thalassaemia up to the 1960s is well exemplified in a series of papers that were presented at a meeting held in New York in 1963, organized by the Medical Advisory Board of the Cooley's Anemia Blood and Research Foundation for Children and the New York Academy of Sciences. Current practice for the transfusion of thalassaemic children was assessed from an enquiry sent to more than twelve centres, caring in all for over 150 patients with severe thalassaemia. The results showed that there were no consistent criteria for determining when a child with thalassaemia should be transfused. Rather, transfusions appeared to be administered either when a particular haemoglobin level was reached or when the children became symptomatic. In hospitals at which blood was administered for symptomatic treatment, the haemoglobin level before transfusion was often as low as 3.0g/dl, and, even if a particular haemoglobin level was chosen, it tended to be somewhere between 5 and 7.0g/dl. It seems very likely that similar information would have been obtained from centres anywhere in the world that were looking after thalassaemic children at this time.

At the same meeting, Irving Wolman of the Children's Hospital of Philadelphia presented the findings of a preliminary study that had set out to determine whether children with severe thalassaemia who had been maintained by blood transfusion at a near-normal haemoglobin level—that is, in excess of 10–11g/dl, were better off than those who had been transfused only when they were symptomatic. Wolman's findings suggested, albeit tentatively, that children maintained at a high—that is, relatively normal—haemoglobin level were taller, had smaller livers and spleens, showed less skeletal deformities and fewer fractures, and had better dental development and less cardiomegaly. In short, it appeared that a high transfusion regimen was able to reverse many of the more distressing features of severe thalassaemia. Wolman was quick to point out that children who had received more transfusions, though more healthy, must have received considerably more iron than those maintained at a lower haemoglobin level. He speculated, however, that, by maintaining a high haemoglobin level, there might be much less stimulus to increased iron absorption from the intestine, so that, in the long term, these children might accumulate less iron than those who grew up with consistently low haemoglobin levels.

At the second meeting of the Cooley's Anemia Blood and Research Foundation for Children, held in New York in 1969, Wolman was able to present follow-up data on the children whom he had described earlier. By then it was clear that those maintained on a high transfusion programme had grown well during their early years, although it appeared that they did not show the usual adolescent growth spurt. But they continued to have less skeletal deformity and bone disease, and their spleens and livers had remained small. However, a note of warning was sounded. Of a total of seventeen children who had received the high transfusion regimen for long enough for adequate evaluation, six had died, all from disorders that could be related to the damaging effects of excess iron in the body.

At the same meeting Wolman's results were confirmed by similar studies from the USA and Europe. By now there was little doubt that children maintained on high-transfusion regimens fared much better than those who were transfused only when symptomatic. It appeared that a high haemoglobin level suppressed ineffective erythropoiesis and appeared to retard the development of enlargement of the spleen and liver and the severe skeletal deformities that had characterized the disease in earlier years.

However, during this period there was growing evidence that children maintained on a high-transfusion regimen were starting to show the side effects of severe iron loading by the time they were entering adolescence, and deaths occurred frequently owing to the damaging effects of iron on the tissues, particularly the heart, liver, and endocrine glands. Clearly, unless it was possible to find a way of removing excess iron from the body, these children would enjoy a much better quality of life in childhood only to succumb to the effects of iron loading before or when they reached adolescence.

Removing Excess Iron: the Development of Iron-Chelating Agents

The reason why the removal of excess iron from the bodies of children with severe thalassaemia presented such a daunting problem, and indeed why they become iron loaded in the first place, is best understood against a background of iron metabolism in normal people, the fundamentals of which were known by the 1960s. Since the body has no way of getting rid of excess iron, the level of iron in health is regulated by the rate that it is absorbed from the diet. Normally, approximately 2mg of iron are absorbed from the diet each day and approximately 0.5–1mg is excreted through the intestine or by loss through the constant shedding of cells from the skin and other tissues. Women in reproductive life lose, on average, an extra 1mg per day through menstrual losses. In normal adults the total amount of iron in the body is approximately 5g, of which 1g is in the storage compartment, 0.5g in the tissues, and the remainder in the circulating red blood cells. For reasons that are not fully understood even today, when there is severe anaemia associated with a great expansion of the erythroid bone marrow, as occurs in thalassaemia, there is an increase in the rate of absorption of iron from the intestine. In addition, if a child is receiving regular transfusion, he or she accumulates a further 200mg of iron for each unit of blood transfused. It is easy to understand, therefore, why a child who has been transfused from early infancy will accumulate 20 or 30g of iron by the time he or she is a teenager. By the 1960s it was clear that the large excess of iron has a particular predilection for the liver, endocrine glands, and, most importantly, the heart.

One of the main problems that faced scientists who were attempting to develop agents that would remove iron—that is, iron-chelating agents—is specificity. While many chemicals can bind iron, they also bind other metals and hence they cannot be used for treating iron-loaded patients because they would deplete them of other vital metals. In the early 1960s the Swiss Federal Institute of Technology in Zurich and the CIBA Pharmaceutical Company (later to become Novartis) in Basle became interested in sideramines, naturally occurring iron-binding agents that are required by bacteria to incorporate iron, which is vital for their growth and development. This led to the development of a potent iron-chelating agent, desferrioxamine, with a sideramine obtained from *Streptomyces pilosus*. Several groups studied the effect of desferrioxamine on iron removal in transfusion-dependent thalassaemic children. At the meeting held under the auspices of the New York Academy of Sciences in New York in 1963, Sephton-Smith reported his experiences with this drug and another chelating agent that has a particular high affinity for iron, diethylenetriaminepentaacetate (DTPA). Although both these agents seemed to be capable of removing iron, neither of them could be administered by mouth, and, because DPTA had to be given by a deep intramuscular injection, mixed with an anaesthetic, whereas injections of desferrioxamine, although still painful, were less traumatic, work over the next few years focused more on this agent.

Several groups in Europe and the USA attempted to treat children with desferrioxamine, but by the late 1960s it appeared that the results of these small trials were disappointing. In most cases it seemed impossible to achieve a negative iron balance, and, because the drug had to be given daily by intramuscular injection, it was poorly tolerated. Indeed, some workers in the field questioned the validity of its use as a chelating agent in the future management of thalassaemia or any form of anaemia associated with excess iron loading.

However, a publication in 1974 from Barry and a team at Great Ormond Street Hospital for Children, London, revived interest in the use of desferrioxamine. This paper described the results of a seven-year trial in which a group of patients were treated with regular intramuscular doses of desferrioxamine together with an intravenous injection at the time of transfusion. A control group was maintained on blood transfusion at a similar level to those who were receiving the drug but received no desferrioxamine or any other chelating agent. When the two groups were analysed after seven years, the serum ferritin level, an approximate though not entirely accurate measurement of body iron levels, was significantly lower in the group that had been treated with desferrioxamine. And, even more importantly, the concentration of iron in the liver in the treated group was lower than in the control group. These children were maintained under observation, and after ten years from the initiation of the trial there was a hint that there might be a decrease in the number of deaths in the treated group, although the numbers were too small to draw any definite conclusions. However, there is no doubt that this study, and the encouraging results obtained by high-transfusion regimens, led, in the mid-1970s, to a major review of the question of the most effective way to administer desferrioxamine.

In his early work on the use of desferrioxamine Sephton-Smith had found that significantly more iron can be removed by continuous intravenous infusion of desferrioxamine, an observation that was later confirmed by others. In 1976 Richard Propper and his colleagues in Boston found that, if they gave an intramuscular dose of desferrioxamine, they removed only 15–26mg of iron; after continuous intravenous administration of the same dose over 24 hours, they were able to remove over 70mg of iron, which is almost the theoretical maximum that the drug can chelate.

While these results were extremely encouraging, it is, of course, impractical to carry out long-term intravenous infusions of drugs in a child's home. Hence Propper and his colleagues examined the pattern of iron excretion after continuous infusions of the drug given subcutaneously with a clockwork pump. This approach, which was also studied by workers in Great Britain, suggested that, although there is marked variability in individual response to desferrioxamine given in this way, a plateau of iron excretion is achieved during the period of infusion and is maintained until the infusion is stopped. The only problem with this technique, of course, is that it would have to have entailed children carrying a pump almost continuously, not something that would be entirely compatible with a normal school life! Based on this concern, Martin Pippard and his colleagues in Oxford found that the drug could be given as a twelve-hour subcutaneous infusion during sleep. It was found that for many patients there was no real advantage in using a longer infusion, because the same amount of iron could be removed by doubling the dose of desferrioxamine over twelve hours as by doubling the infusion time. A number of groups found that, during these studies, an even larger amount of iron could be removed if the children received ascorbic acid (vitamin C) at the time of the infusion.

Using these regimens, together with the development of improved pumps for the slow infusion of desferrioxamine, genuine progress was made towards controlling iron loading in transfusion-dependent children with thalassaemia, at least in countries that could afford this expensive form of treatment. Remarkably, long-term experience with the use of desferrioxamine suggested that it was relatively free of toxic effects, with the exception of occasional complications affecting the eyes, hearing, or growth, which could be avoided by careful monitoring. But, of course, tissue damage from iron loading takes many years to evolve, and it was not until the mid-1990s that the classical studies of Nancy Olivieri, Garry Brittenham, and others showed beyond any doubt that, provided that the ferritin levels in the blood or iron levels in the liver can be maintained at clearly defined levels, patients with severe thalassaemia treated in this way grow and develop well and can be maintained in good health over a long period.

This was not the end of the problem of iron loading, however. Having to go to bed at night with a subcutaneous needle inserted into the abdominal wall is not easy for either the patient or his or her parents, and poor compliance became a problem that has dogged the use of desferrioxamine up to the present day. This critical issue led to a search for an effective chelating agent that could be taken by mouth, a topic to which we will return in a later chapter.

Splenectomy

The question of the role of removal of the spleen for the treatment of thalassaemia has been long-standing and highly controversial. There were many reports of splenectomy for the treatment of splenic anaemia, or Von Jaksch's anaemia, in the early part of the twentieth century, before Cooley's and Lee's first description of a severe form of thalassaemia. The practice continued, although by 1936, in her monograph on anaemia, Janet Vaughan concluded that the operation was of no value for the treatment of children with this disease. By the early 1960s a completely confusing literature had amassed around this subject, but, overall, most studies seemed to demonstrate that at least a proportion of children with severe thalassaemia might have a reduction in transfusion requirements after the operation.

What was, and still is, the rationale for removing these enlarged spleens? In some cases they reach a size such that they cause considerable discomfort and pain. Furthermore, as it became possible to study the distribution of red blood cells by using radioactive isotopes, it became apparent that these very large spleens can trap a considerable proportion of the circulating red cells and also cause their increased rate of destruction. Also, and for reasons that are still not clear to this day, massive spleen enlargement or splenomegaly is often associated with an increased plasma volume, causing further worsening of the anaemia. And it became clear that, at least in some children with thalassaemia, removal of a large spleen is followed by an improvement in the rate of growth.

When better blood-transfusion regimens were introduced in the early 1960s, it was found that enlargement of the spleen was less common, but the problem still remained. It was not until the mid-1970s that the role of splenectomy became better defined. This was largely through the work in London of Bernadette Modell and her colleagues, who attempted to anticipate the development of hypersplenism—that is, trapping of the red cells and their increased destruction in the spleen by drawing a series of standard curves for monitoring haemoglobin levels after blood transfusions. From her studies, it appeared that the rate of fall of haemoglobin might be a useful guide to whether a splenectomy would be helpful; of fifty-eight splenectomized patients studied in this way, all but three had a permanent reduction in blood requirement after the operation.

Although removal of the spleen is a fairly safe surgical procedure, it had been realized since the early 1940s that children who had lost their spleens were more prone to infection. In the early 1960s Carl Smith and his colleagues in New York produced extensive evidence that post-splenectomy infection was a frequent and often disastrous complication in thalassaemic children. Since this complication was particularly common if the spleen had been removed in early childhood, it became common practice to withhold the operation until children had reached at least the age of 5 years. Later, prophylactic penicillin and vaccination programmes were introduced in an attempt to reduce the frequency of overwhelming infections in these children.

But, as more experience was obtained of maintaining the haemoglobin level of children with severe thalassaemia at relatively normal levels by adequate transfusion, it became apparent, during the late 1970s and after, that enlargement of the spleen was a clear indication of an inadequate transfusion programme, and this realization led to a marked reduction in the requirement for splenectomy in subsequent years.

Improved Symptomatic Treatment, but Problems Remain

The development of more adequate transfusion regimens together with chelating agents undoubtedly transformed the lives of children with severe thalassaemia during the 1960s and 1970s. In richer countries, in which children could be managed by these approaches, their quality of life improved immeasurably, and the hideous bone changes that involved the face and skull ceased to be a problem. Up until the early 1960s textbooks on the management of thalassaemia included sections on the surgical treatment of the complications of skull deformity such as deafness and involvement of the optic nerves, not to mention articles on how to manage the serious dental deformities of these children. Within ten years these accounts were becoming a thing of the past.

But problems remained. Compliance to the demanding overnight desferrioxamine infusions was far from perfect, and growth disorders and deaths that were due to cardiac iron deposition continued, though at a much lower level. In the developing countries, particularly those of Asia, where apart from occasional blood transfusion there were no other forms of treatment available, the disease continued to take its toll as before. And in many countries there was an increasing problem owing to blood-borne infection, notably with hepatitis B and C, malaria, and, later, HIV/AIDS.

Of course, all this treatment was symptomatic and had to continue throughout a patient's life. It was not until the first bone marrow transplantations were carried out in 1982 that a definitive cure for the disease was achieved, and even then only for a limited number of patients, a topic to which we will return in a later chapter.

Can a Genetic Disease Like Thalassaemia be Avoided?

Considering the inconvenience and discomfort of the treatment that was available for thalassaemia in the 1970s, it is not surprising that thoughts turned towards the possibility of controlling the numbers of babies born with this condition, particularly in countries in which there was a very high frequency of the disease. In this context I have used the word avoidance rather than prevention. It is not possible to 'prevent' genetic disease, because mistakes during DNA copying are inevitable; the wonder is that they do not happen more often.

The social and economic consequences of the high frequency of thalassaemia were graphically illustrated in Cyprus, a country that underwent an epidemiological transition after the Second World War following a major malaria eradication programme and accompanying improvements in public-health measures. It soon became clear that there was a very high frequency of babies with severe thalassaemia on the island, and by the early 1970s it was estimated that, if no steps were taken to control the frequency of the disease, in about forty years time the blood required to treat all the severely affected children would amount to some 78,000 units per annum, 40 per cent of the population would need to be blood donors, and the total cost to the island's health services would equal or exceed its health budget.

There are two ways in which a reduction in the numbers of babies born with severe thalassaemia might be achieved. Since by the mid-1960s or even earlier it was possible to identify the carrier states for different forms of thalassaemia, it might be feasible to screen populations and to offer marital advice to those who were found to be carrying a thalassaemia gene. The other way forward would be to try to find a way of identifying whether a fetus had received a thalassaemia gene from both parents and then to offer the couple termination of the pregnancy if this were the case. This would require screening in the ante-natal clinic and, if the mother was found to be a carrier, screening her husband to see if they were at risk of having a severely affected child.

The earliest population screening programmes for inherited haemoglobin disorders, initiated in the early 1970s, focused mainly on sickle-cell anaemia and were not a resounding success. The Afro-American population in the USA was informed that many of them were suffering from this neglected disease, and a large-scale screening programme was established, backed up by heavy federal support. Local communities became involved, together with an ill-assorted band of individuals of diverse background and training; the programme soon became a major political, racial, and social issue. Many of the screening programmes were not backed up with appropriate provision for genetic counselling; in several states laws were passed that made screening mandatory without any provision for education or counselling. This ill-directed activity caused large-scale public anxiety, stigmatism, job and health-insurance discrimination, and many other undesirable effects, and virtually nothing was achieved.

A much better-conceived and organized genetic screening programme was carried out in the early 1970s in a farming community, Orchomenos, 150km north of Athens, in which a high incidence of sickle-cell disease was already well recognized by the community. The programme had been following goals: education of the community about the genetic facts of sickle-cell anaemia, testing for the sickle-cell trait in families in the community, an explanation about the significance of a positive test with precise definitions of the options and risks, and an introduction into the local culture of a campaign designed to ensure that prospective spouses exchanged information about their haemoglobin genotypes before marriage. The programme, which was backed up by small centres staffed by two physicians dealing with areas containing up to 2,300 families, lasted for 3½ years, during which careful follow-up studies were carried out regularly. Almost all the families interviewed remembered the results of their blood tests and the associated genetic implications; there seems little doubt that the major target of the programme—to introduce the idea of premarital exchange of genetic information into the culture—was achieved. However, it is equally clear that the programme caused serious problems in the community and led to many social stresses and strains. People who carried the sickle-cell gene were unwilling to talk about their genetic make-up, and, because there were many arranged marriages in the community, potential marriage partners had to live in a state of uncertainty up to the moment when the haemoglobin status of the future in-law was revealed. Much of the anxiety generated seemed to be the result of social stigmatism subsequent to the collapse of marital arrangements because of the sickle-cell trait. In some cases individuals simply kept quiet about the fact that they were carriers.

In the event, the overall outcome of the Greek programme was disappointing. If mating had been completely random in the population, the expected number between heterozygotes would have been 4.5 per cent. In fact, in the 101 matings studied during the programme there were 4 such matings. Interviews of the spouses showed that in two cases one of them had concealed her carrier status; in the other two the couples were married with a complete understanding of the risks. This small but carefully carried-out study suggested that, even in a community whose members are well aware of the risks of a particular genetic disease, and who are subjected to a carefully designed screening programme backed up by adequate counselling, it is extremely difficult to achieve anything more than a random mating pattern. While Greeks are notorious for being an independently minded people it is doubtful whether this outcome would differ in other populations!

Another extensive programme of this type was established for schoolchildren of intermediate grades, organized by Silvestroni and Bienco at the Centro di Studi Della

Microcitemia di Roma. The organizers established the programme on the basis that it could be cost effective, because in Latium there were about 150 patients with homozygous β thalassaemia and the annual cost of screening in schools over the whole region would equal that of the annual medical costs of maintaining 20–25 such patients. The Italian workers carried out careful follow-up studies and stated that they had not created anxiety or tensions in the thalassaemic families and that they were confident that the affected children would be prepared to take their genotype into account when they were about to marry. A later telephone survey suggested that even after four years the families with thalassaemic children had retained the gist of the genetic information. Unfortunately, however, the long-term outcome of this extensive study was never published, and so we will never know its potential effect on the birth of affected babies.

In 2002 Canali and Corbellini published an extensive review of the early evolution of control programmes for thalassaemia in Italy, which were not confined to the programme in Rome but involved several centres in other parts of Italy. They ascribe its lack of success to the complex interplay of many social and political factors, including isolation and lack of funding for the centres involved, poor coordination, and, above all, inadequate education of both medical staff and the population. They add, however, that it did form the basis, at least in some centres, for the later success of thalassaemia control in Italy.

Considering these disappointing results, it is not surprising that in the early 1970s thoughts turned to whether it might be possible to identify the disease in affected fetuses and to offer termination of pregnancy.

Prenatal Diagnosis

Prenatal diagnosis for genetic disease was pioneered in the early 1950s for assessing the level of bilirubin in amniotic fluid for the diagnosis of rhesus haemolytic disease, and was further adapted in 1956 for determining the sex of the fetus. By the mid-1960s it had been found that amniotic fluid cells could be cultured, allowing both chromosomal and biochemical analysis for the diagnosis of different genetic disorders.

The main stimulus to the notion that it might be possible to identify common haemoglobin disorders in the fetus came from studies of the haemoglobin composition of fetal blood and the pattern of globin synthesis during early fetal life that were carried out between the mid-1950s and the early 1970s. As discussed in previous chapters, the main type of haemoglobin that is produced during fetal life is Hb F, which has the structure α_2, γ_2. Late in pregnancy and in the early months after birth Hb F is replaced by adult haemoglobin, Hb A, which has the structure α_2, β_2. In other words, the switch from fetal to adult haemoglobin production reflects a change from γ chain to β chain synthesis. The earliest electrophoretic studies of fetal blood showed that there were small amounts of adult haemoglobin present quite early during fetal development. Later, *in vitro* studies of haemoglobin synthesis showed quite unequivocally that β chain synthesis is activated at about the eighth week or even earlier and reaches a steady state level of about 10 per cent of that of γ chains up to about 30–40 weeks gestation. Furthermore, the relative rates of β chain and γ chain synthesis are synchronized throughout the different organs of blood production in the fetus. These observations suggested that, if small blood samples could be obtained from fetuses between 20 and 25 weeks gestation and subjected to globin-synthesis analysis, which was described in an earlier chapter, it should be possible to diagnose homozygous β thalassaemia *in utero*, assuming of course that the β thalassaemia mutation is expressed early in fetal life. Early studies on aborted fetuses that were at risk of being homozygous for β thalassaemia showed that this is the case. Indeed, as early as 1972 this approach was used successfully for the prenatal diagnosis of sickle-cell anaemia.

Perhaps because it was my own team that had developed the *in vitro* globin chain synthesis method, we were at first sceptical about whether it would be sufficiently sensitive to identify defective β chain synthesis in fetal life. However, several groups decided to explore this approach, including those led by Nathan, Kan, and Alter in the USA, Modell and Rodeck in England, Loukopoulos in Greece, Angastiniotis in Cyprus, Vullo in Italy, and Cao in Sardinia.

The major difficulty, of course, was how to obtain fetal blood samples for analysis by globin synthesis. In the event, two ways were explored, placental aspiration and direct aspiration of samples from placental vessels by fetoscopy. Not surprisingly, many of the blood samples obtained by these methods contained mixtures of fetal and maternal red cells. Not daunted, several workers explored ways of separating them, or at least trying to calculate the degree of contamination of one by the other. Although these techniques were used in some of the early attempts at prenatal diagnosis, they proved difficult, and hence thoughts turned to direct fetal blood sampling. As the technique of fetoscopy improved, it became possible to obtain fetal blood directly, an advance that greatly facilitated the prenatal diagnosis of thalassaemia during the mid-1970s. The problem of contamination of fetal blood samples with small amounts of maternal blood remained, however. In the end it was solved by using methods based on the difference in size between fetal and adult red cells, together with staining methods for identifying the relative numbers of maternal cells in fetal blood samples.

By the end of 1978 teams of workers in the USA, England, Sardinia, Cyprus, Greece, and Italy were able to publish the results of these early attempts at the prenatal diagnosis of β thalassaemia. Based on experience of well over 100 attempts, the technique seemed to be feasible for identifying homozygous β thalassaemia and sickle-cell anaemia; there had been remarkably few diagnostic errors, and the only worrying feature was the relatively high fetal loss attributable to the procedure. By 1989, 13,921 prenatal diagnoses had been carried out in twenty centres in different countries using fetal blood sampling. The effect on the birth rates of new cases of severe thalassaemia in some countries was quite dramatic (Figure 12).

This was a particularly pleasing episode in the story of the exploration of the thalassaemias and an elegant example of how curiosity-driven research can rapidly find application in the clinic. For, in little over five years, an experimental technique that had been developed to explore the rates of globin chain synthesis, and hence the pathogenesis of the disease, had been applied in several countries for its effective prenatal diagnosis. It was to be another ten years before this approach was largely superseded by fetal DNA analysis, a

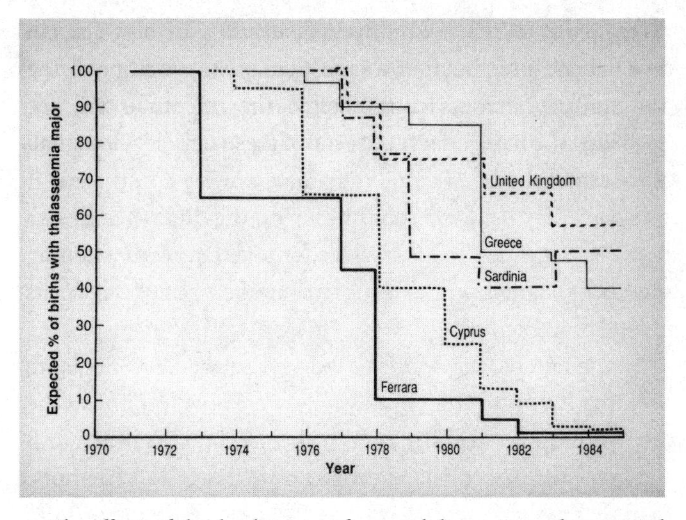

12. The effects of the development of prenatal diagnosis on the expected number of births with thalassaemia. From 1972 until about 1982 these results were obtained using globin chain synthesis of fetal blood, after which most centres exchanged this method for fetal DNA analysis. (*Adapted from Modell and Bulyjenkov (1988).*)

time during which *in vitro* globin biosynthesis was applied widely in many populations and led to a major decline in the number of births of infants with severe forms of thalassaemia, particularly in very high-frequency countries like Cyprus and Sardinia.

These approaches to the avoidance of thalassaemia raised several ethical issues, which were hotly debated during the 1970s, an important topic to which we shall return in a later chapter.

Hence, by the end of 1970 the basis for the avoidance and better management of the thalassaemias had been firmly established. We shall return to the later developments that were further to improve the lives of patients with these diseases in a later chapter, but first we must complete the story of how the study of thalassaemia entered the molecular era and provided the basis on which some of these new developments were established.

VI

THALASSAEMIA AND THE DAWN OF MOLECULAR MEDICINE

Probably because it is too soon, the history of the early development of what is now called molecular medicine has not yet been recorded. When historians come to focus on this remarkable story, they are likely to be impressed with how often it was that key discoveries relied on the development of new technology rather than original ideas, a phenomenon common to many of the biological sciences in recent times. But, before describing the major role that the study of the thalassaemias played in the early development of what, sometimes rather hopefully, became known as molecular medicine, it may be helpful to readers who are not familiar with the basics of molecular biology to outline where this field had reached by the early 1970s.

The Background to Molecular Medicine

By the early 1970s it was well established that genes consist of DNA molecules, which are made up of two chains of nucleotide bases wrapped round each other. There are four bases, adenine (A), guanine (G), cytosine (C), and thymine (T).

Essentially, a gene consists of a variable length of these nucleotides and contains the information required to put amino acids in the appropriate place in a peptide chain together with crucial information about how its activity should be controlled. The genetic code—that is, the information required correctly to order the amino acids in a peptide chain—is a triplet code consisting of three bases. The extraordinary facility of DNA as a self-copying machine is due to the rules of pairing between the bases; A can pair only with T, and C with G. Therefore, when the two strands of DNA divide, each can act as a template only for a new strand that has exactly the same base-pair composition.

Genes, of course, are found in the nucleus of our cells, whereas proteins are synthesized in the cytoplasm—that is, the region outside the nucleus. Thus, when a gene is appropriately activated as a prelude to producing its protein product, one of the DNA strands acts as a template for the production of a ribonucleic acid (RNA) molecule, messenger RNA (mRNA), which, with one exception, has the same bases as DNA; instead of thymine, mRNA contains the closely related base, uracil (U). Hence, when mRNA is synthesized on its DNA template, it carries a complementary message to the cell cytoplasm. Individual amino acids are transported to the mRNA template by another class of RNA molecules called transfer RNAs, each of which is specific for a particular amino acid. The first amino acid is put in place at the appropriate start point, or initiation codon, and then the second amino acid moves in alongside and the two are bonded together. This sequential process continues until the stop point, or termination codon, on the mRNA template. As mentioned in a previous chapter, the growing chains are held in place by bodies called ribosomes, and when the process is complete the growing chains and ribosomes fall off the mRNA template. For proteins like haemoglobin with more than one peptide chain, the chains then combine to form the completed molecule.

Up until the 1970s the central dogma of gene action was that there was a direct relationship between the codons of DNA with those of mRNA and the amino acids that were being inserted into a growing peptide chain. This concept of the continuity between gene and protein structure was delivered a major blow in 1977 with the discovery that the sequences of genes that direct the order of amino acids in proteins are interrupted by variably sized regions of nucleotides, which seemed at the time to have no informational value whatever, a finding that those who held the central dogma close to their heart found extremely depressing. Francis Crick, the co-discoverer of the structure of DNA, described his reaction to the news as like reading an elegant paragraph of prose by Jane Austen, only to come across a deodorant advertisement in the middle! But all mammalian genes seemed to be broken up in this way; the coding regions were called exons and the non-coding regions introns.

Of course, the discovery of introns required that there be a further critical step in the transference of genetic information from the cell nucleus to the cytoplasm. After its synthesis on its DNA template, mRNA contains both exon and intron sequences. Hence, while still in the nucleus, the intron sequences have to be cut out of the mRNA and the exons joined together, or spliced, before the definitive mRNA is transported to the cytoplasm to act as a template for protein synthesis. Although it took many years for the details of processing of mRNA precursors to be worked out, this new step in gene action also offered a variety of opportunities for things to go wrong and for the production of genetic disease.

Although this basic information, which is summarized in Figure 13 in a simplified form, is all that is needed to understand the objectives of those who started to explore the molecular

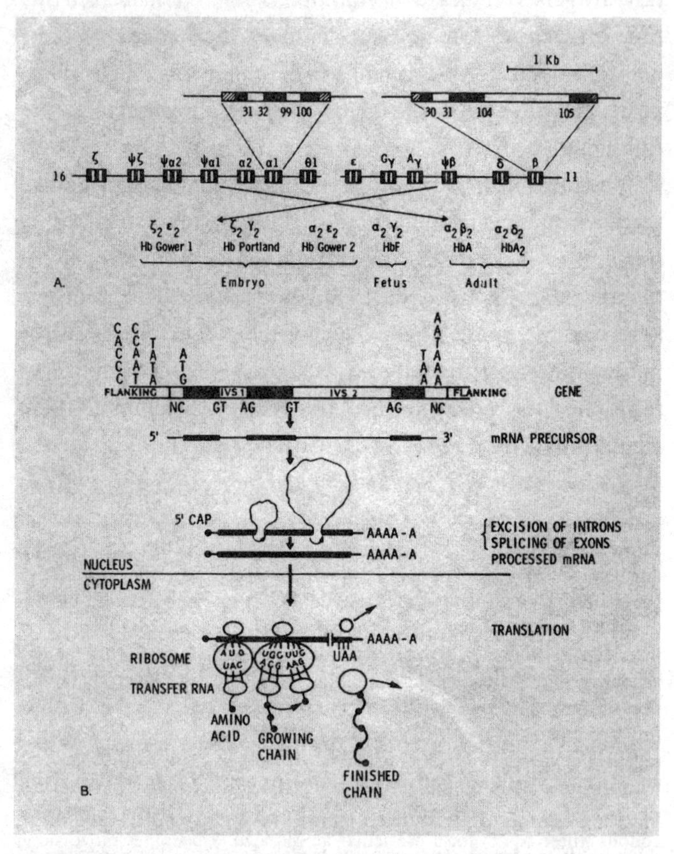

13. The genetic control and mechanisms of haemoglobin synthesis. A. The α-like and β-like globin gene clusters on chromosomes 16 and 11 are shown.

basis of disease, it only summarizes the early stages of an enormous explosion of knowledge in biology, the results of which are still evolving and look likely to do so for the foreseeable future: how are genes controlled, why are different genes switched on and off in different cells of the body, why are genes switched on and off at different stages of development, and how are genes regulated in a multi-cellular organism such that they are able to interact one with another? Many of these questions still remain unanswered, and it may take many years, and the help of complex systems biology, to try to put this jigsaw together.

When medical historians of the future study the origins of the development of molecular medicine, and particularly of those who moved the thalassaemia field into the realms of molecular biology in the late 1970s and early 1980s, they will surely be interested in the backgrounds and training of the major pioneers in this field. Surprisingly, with a very few exceptions, they were not molecular biologists. Rather, most of them came from backgrounds in haematology but had managed to

---

Below them are also shown the different haemoglobins that are produced during various phases of human development. At the top of the diagram two genes have been expanded to show the exons in black and the introns unshaded; the number of the base pairs between which the introns are inserted are also shown. Kb represents 1,000 base pairs. B. At the top is shown a globin gene and at either end of the gene the sequences involved in regulation of the gene are included. This is transcribed into a messenger RNA (mRNA) precursor and the introns have to be excised and the exons joined together before the mRNA is delivered to the cell cytoplasm to act as a template for protein synthesis. The growing peptide chain is carried by the ribosomes, and the appropriate amino acids are brought to the appropriate codons (three bases) for a particular amino acid. Ribosomes move along the mRNA until they reach a codon, UAA, which reads 'stop', at which time the completed chain is released together with the ribosome subunits.

pick up enough knowledge about genetics and the evolving field of molecular genetics to be able to apply the new techniques that were continually arising from these fields, or at least have enough knowledge to ask the right questions of molecular biologists at each stage in their research.

In the USA many of the early workers in this field came from Nathan's remarkable stable at the Children's Hospital in Boston, including Kan, Orkin, Forget, Benz, and others. Other leading American groups included those of Bank and colleagues in New York, Nienhuis and Anderson in Washington, Stamatoyannopoulos and colleagues in Seattle, Antonarakis, who also spent time in Boston, and Kazazian in Baltimore. My own team in Oxford, including Clegg, Wood, Higgs, Old, and Hunt, was associated at the beginning of this enterprise with Williamson's group at St Mary's Hospital, London, and later in Glasgow, and regular meetings and workshops were established at which Jeffreys and Grosveld, two of the leading molecular biologists in the field, attended. Especially in England, during these early days in the development of this field, there were very few places where young clinicians could be trained in the technology of molecular biology and, similarly, equally few laboratories where young Ph.D. graduates with a background in this field could work in an environment where its medical applications were being explored. As a potential solution to this problem, the Institute of Molecular Medicine was established in Oxford in the late 1980s.

The molecular analysis of the genetic defects that underlie thalassaemia went through two distinct phases. Between 1970 and 1980 it focused largely on messenger RNA. At first this work was directed at the function and relative levels of mRNA in different forms of thalassaemia, but later, when it became possible to synthesize a DNA copy using an mRNA template, the resulting copy called complementary DNA (cDNA) was used to probe the genome to determine if some forms of thalassaemia might result from the deletion, or loss, of a particular globin gene.

Although our story now moves from the protein products of genes to the genes themselves, it is important to appreciate that studies of normal and abnormal haemoglobins and their genetic control continued through the 1970s. This work culminated in a clear picture of the genetic control of haemoglobin, a vital basis for interpreting the results of DNA analysis that followed. The genetic control of haemoglobin and steps involved in the expression of a globin gene are summarized in Figure 13.

Early Studies of the Function and Quantification of Globin Messenger RNA

In the early 1970s it became possible to analyse the function of messenger RNA in cell-free systems—that is, the contents of cells from different sources that retain the property of synthesizing proteins when appropriate mRNAs are added. The first experiments of this type were carried out in 1971 by Nienhuis and Anderson in Washington and Benz and Forget in Boston. Both of these groups found that, whereas mRNA isolated from normal human reticulocytes directed the synthesis of equal amounts of α and β chains, mRNA from reticulocytes of patients with β thalassaemia made fewer β chains than α chains. Over subsequent years these results were confirmed by several workers.

While these studies undoubtedly showed that there is a deficiency of functional β globin mRNA in some β thalassaemic red cells, because these cell-free systems measured activity and not quantity, they were unable to clarify whether the reduction in biological activity of the added thalassaemic mRNA was due to it being present in decreased amounts or in normal amounts but with defective function. It was quite possible, for example, that some β thalassaemic cells contain completely normal levels of β globin mRNA but with a structural defect that makes it impossible for this to function normally in protein synthesis. The experiments that would distinguish between these two possibilities soon became possible by a further technical advance in the field.

In 1970 David Baltimore and Howard Temin quite independently discovered an enzyme in certain tumour viruses that they called reverse transcriptase, a finding that made it possible to synthesize complementary DNA (cDNA) from mRNA templates, work for which they later received the Nobel Prize. By obtaining RNA from reticulocytes—that is, young red cells that retain the capacity for synthesizing haemoglobin and that therefore contain α and β globin mRNA—and by adding appropriate radioactive nucleotides, it was possible to synthesize radioactively labelled α and β globin cDNAs. Since, because of the rules of base pairing, these could bind only to appropriate α or β globin mRNA, they could be used to analyse the levels of the different forms of mRNA in the reticulocytes of patients with thalassaemia by a process called molecular hybridization. In very early experiments of this type, cDNAs were made from rabbit mRNA templates and relied on cross hybridization between rabbit and human nucleic acid sequences for their success. Later, highly purified cDNAs specific for human α, β, and γ globin mRNA sequences became available.

The first studies using these new tools were carried out on the blood of patients with severe β thalassaemia. The red cells showed a reduced amount of adult haemoglobin (Hb A)—that is, they were synthesizing some normal β globin chains but at a reduced rate. Overall, these early experiments suggested that in these cases there is a reduced amount of β globin mRNA at a level that is more or less in agreement with the overall degree of globin chain imbalance as measured by *in vitro* synthesis studies. The attention of many groups then turned to an analysis of β globin mRNA in the red cells of patients with severe β thalassaemia who were making no Hb A—that is, there was an absence of β chain synthesis. This work, which was carried on throughout the 1970s, demonstrated that, in some cases no β globin mRNA was detectable, while in others there were significant amounts of apparently full-length but non-functional β globin mRNA.

By now it seemed likely that an analysis of the structure of the non-functional β globin mRNA found in at least some patients with β thalassaemia might lead to a genuine understanding of the underlying cause of at least one form of the disease. Using cDNA probes that had been made specifically for RNA sequences at one or other end of the β globin mRNA, Old and his colleagues in Oxford were able to demonstrate, although not absolutely prove, that there might be a defect in or near the initiation site for globin chain synthesis in one particular case and, in another, a deletion or insertion of abnormal RNA at the other end of the abnormal globin mRNA. These observations were extended by Kan and his colleagues in San Francisco, who determined the nucleotide sequence of non-functional β globin mRNA from a Chinese patient with β^{o} thalassaemia. They found that the AAG codon for lysine at position β 17 had changed, or mutated, to the chain termination codon UAG, thus leading to premature β globin chain termination, with the production of a short, 15-amino-acid-length fragment.

In the late 1970s two groups of workers observed that there were substantial amounts of β globin-like RNA sequences in the nucleus of the red cells of some patients with β thalassaemia, but not in the cytoplasm of the red-cell precursors. These findings suggested that there might be yet other mutations that produced a defect in processing, or, alternatively, resulted in extreme instability of β globin mRNA.

Thus the 1970s were a period of enormous productivity that yielded increasing insight into the remarkable genetic heterogeneity of the β thalassaemias and, for the first time, suggested that some cases might result from deletions of parts of the β globin gene while in others point mutations—that is, single nucleotide replacements—might generate β globin mRNA that is non-functional.

mRNA and α Thalassaemia

Readers will recall that by the beginning of the 1970s it was clear that there must be at least two forms of α thalassaemia, α thalassaemia 1, in which few or no α globin chains are produced, and α thalassaemia 2, in which there is a mild reduction in the production of α chains. These two conditions were later renamed α^0 thalassaemia and α^+ thalassaemia, respectively, to keep in line with the nomenclature for the different forms of β thalassaemia.

Studies of mRNA in the α thalassaemias during the 1970s followed exactly the same pattern as those for the β thalassaemias. Analysis of the relative amounts of α and β globin mRNA in cell-free systems demonstrated that there is a deficiency of functional α-globin mRNA in patients with Hb H disease. These findings were confirmed by cDNA hybridization studies, and it was found that there is no α globin mRNA in babies with the Hb Bart's hydrops syndrome, mirroring the complete absence of α globin chain production in the blood of these babies.

By 1980 the purity of the cDNA probes and the sensitivity of these techniques were developed to such a degree that it was possible to show a clear separation of levels of α globin mRNA between normal people, α^+ thalassaemia or α^0 thalassaemia carriers, and patients with Hb H disease who had inherited both defective α globin genes from their parents.

In 1970 there had been a further twist to the story of a possible defect in α globin mRNA. Paul Milner wrote to us and told us about a Chinese family in Jamaica in whom there were patients with Hb H disease who had a very small amount of an abnormal haemoglobin demonstrable by electrophoresis. A family study suggested that these patients had inherited a severe α thalassaemia gene from one parent and a gene for this highly unusual haemoglobin variant from the other. When the structure of the variant, named Hb Constant Spring after the name of the suburb of Kingston where the family lived, was analysed, we found that it consisted of normal β chains combined with α chains that were extended by thirty-one additional amino acids from beyond the natural terminal arginine amino acid (codon 141). By then it was already known that the normal termination codon in mRNA, which signals protein synthesis to cease, has the structure UAA. Because the first amino acid in the extended α globin chain was glutamine, the code word for which is CAA, we suggested that the termination codon UAA had changed to CAA and hence the mRNA continued to be translated until another termination codon was reached. At that time, the structure of mRNA beyond the normal termination codon was not known, but from the amino-acid sequence of the extended region of the abnormal α chain it was possible to deduce what this might be, an observation that was treated with some scepticism by molecular biologists who were sequencing this region of α globin mRNA at the time. They later begged forgiveness when the predicted sequence was found to be absolutely correct!

Since the α globin gene termination codon could theoretically mutate to a codon that would code for several other different amino acids, it was predicted at the time that more elongated α globin chain variants would be discovered with different amino acids at the beginning of the elongated sequence; within a few years several were found. Clearly, therefore, at least some forms of α thalassaemia were due to a mutation in the termination codon for globin chain synthesis, probably reflecting instability of the abnormally translated α globin mRNA.

Probing the Globin Genes in Thalassaemia

Since by the early 1970s it was already clear that at least in some forms of thalassaemia there was an absence of α or β globin chain synthesis, it was possible that the particular genes that regulate their production might be missing, or deleted. Now that complementary DNA probes that would anneal, or hybridize, to α or β globin mRNA or to DNA of the same sequence were available, it was postulated that it should be possible to use these probes to determine whether the α or β globin genes were present in the DNA of individuals with α or β thalassaemia.

In the early 1970s we approached John Paul and his group in Glasgow, who already had considerable experience of this type of molecular hybridization, to ask him whether an experiment of this kind would be feasible. The idea was to obtain DNA from the tissues of a stillborn baby with the Hb Bart's hydrops syndrome in whom we had found absent α chain synthesis and compare the pattern of cDNA hybridization using an α globin gene probe with that of a stillborn infant with no evidence of thalassaemia (Figure 14). The experiments, carried out by Ottolenghi and Williamson in Paul's laboratory, showed that DNA obtained from the liver of the baby with hydrops contained normal amounts of β globin DNA but a more or less complete absence of α globin DNA, while in DNA prepared from the same tissues of normal infants the expected α and β DNA sequences were present. This result showed beyond any reasonable doubt that the severe form of α thalassaemia that the hydropic infant had inherited from both parents was due to a major deletion of both the α globin genes. A paper describing these results in *Nature* appeared back to back with a paper describing an identical study carried out quite independently by Kan and his colleagues in San Francisco at about the same time. These studies showed for the first time that genetic diseases can result from the partial or complete loss, or deletion, of a particular gene, or genes.

These findings were further augmented by experiments carried out by Kan's team that suggested that the hybridization pattern of DNA from a patient with Hb H disease, the milder but symptomatic form of α thalassaemia that is compatible

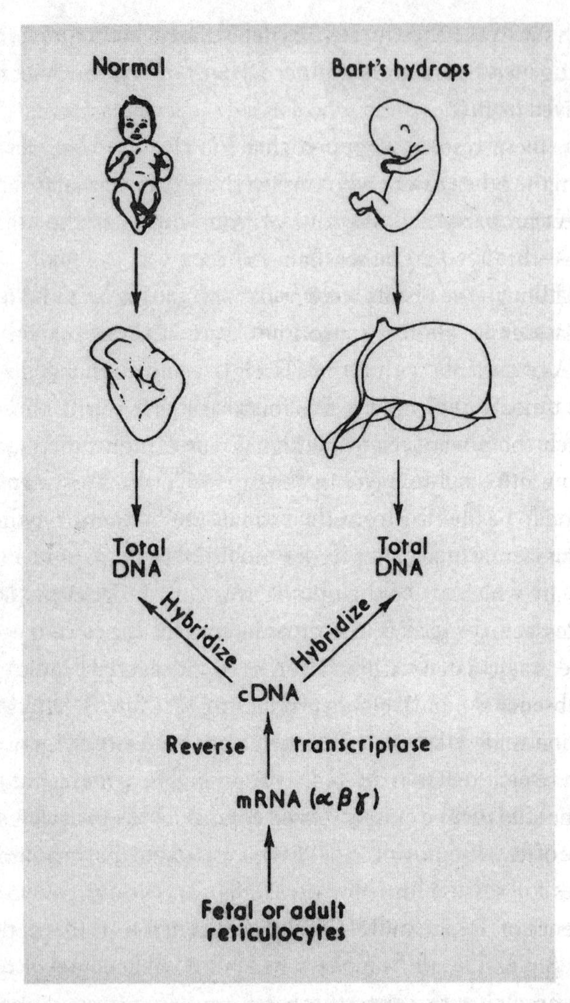

14. The demonstration of a complete deletion of the α globin genes in babies with the Hb Bart's hydrops syndrome. mRNA was obtained from normal fetal or adult blood and transcribed into complementary DNA (cDNA), which was hybridized to DNA obtained from either normal fetal tissue or tissue from a baby who had died with Hb Bart's hydrops. The experiment demonstrated an almost complete deletion of the α globin genes in the affected baby.

with survival, has only one-quarter of the normal complement of α genes. Since there are no α genes on the chromosome derived from the parent who carried the α° thalassaemia deletion, these results suggested that the chromosome derived from the other parent who carried the milder α^{+} thalassaemia defect probably had only one α gene, instead of the normal two—that is, α/αα rather than αα/αα.

Although the results were not as spectacular as those for α thalassaemia, similar approaches were used to analyse the DNA of patients with β thalassaemia and its variants during this time. For example, it was found that, in some of those in which there were relatively high levels of non-functional β globin mRNA, the β globin genes were intact. This was also found to be the case for at least some cases of homozygous β° thalassaemia in which there was no β globin chain production and in which β globin mRNA was largely lacking. In δβ thalassaemia—that is, the disorder that in the homozygous state is associated with no Hb A or A_2 production, indicating an absence of δ and β globin production—cDNA/DNA hybridization analysis suggested that at least 75 per cent of the normal β and δ gene region is deleted. By applying the same techniques to the study of the closely related condition hereditary persistence of fetal haemoglobin (HPFH), it was found that in homozygotes for at least some forms of this condition there was an absence of β globin mRNA and that at least two-thirds of the δ and β globin gene sequences had been deleted. However, it appeared that there were other forms of this condition in which the β and δ globin genes remain intact.

Considering that the extensive molecular hybridization studies of mRNA and DNA carried out throughout the 1970s were all, in effect, indirect approaches to the analysis of the genetic defects that underlie the thalassaemias, it was remarkable how much had been learnt. By the late 1970s it was known that the common severe form of α thalassaemia is due to the deletion of both of the linked pairs of α globin genes and that, almost certainly, the milder form results from the loss of one of the pair. Furthermore, there is another form of α thalassaemia that turned out to be common in Asian populations that is due to a mutation in the chain-termination codon of one of the α genes. It was also becoming clear that the β thalassaemias are highly heterogeneous at the molecular level, and that at least a few of them result from gene deletions; in one case the disease is caused by a mutation that produces a premature chain-termination codon. It was also established that the δβ thalassaemias and some forms of HPFH are due to deletions of different sizes that involve the β and δ globin genes.

While all this work was being carried out, quite remarkable advances were occurring in the technology of molecular biology, which were to make it possible to move towards the final step in the analysis of the thalassaemias and to explore their defective genes directly.

The Technical Revolution Continues

Considering the extraordinary pace and breadth of the different techniques for analysing DNA that appeared in the 1970s, it is impossible to discuss this remarkable period in the evolution of molecular biology in detail, but if readers are to understand how the analysis of the genetic defects in thalassaemia was to pave the way for the development of what became known as molecular medicine, it is important that they have a broad understanding of how this remarkable new technology was able to move the field forward.

One of the key advances was the construction of more specific probes for identifying changes in DNA. The probes that were described in the previous sections were relatively large molecules that were able to identify individual genes or parts of them. In other words, they bound, or hybridized, with long stretches of complementary sequences, but they were not capable of recognizing single base changes or similar small structural alterations. This problem was solved by making use of short, synthetic DNA fragments called oligonucleotides, which could be constructed specifically to detect single base changes in DNA.

Another seminal advance was the discovery of a family of bacterial enzymes called restriction endonucleases, so named because of their ability to identify foreign DNA in the cells of bacteria. A large family of these enzymes was isolated and purified, with each enzyme named after the particular bacterium from which it was isolated. These enzymes do not cleave DNA at random, but do so only where a particular sequence of bases occurs in the molecule. Those used most commonly in genetic engineering recognize signals consisting of four, five, or six bases.

With the discovery of restriction enzymes, it became possible to embark on restriction endonuclease mapping, or gene mapping, which played a central role in the early analysis of genetic disease at the DNA level. Indeed, it is impossible to overemphasize its importance in the development of human molecular genetics. In short, DNA can be obtained from any available tissue and, after purification, digested with a particular enzyme. The mixture of the different-sized fragments is then subjected to electrophoresis, after which the separated fragments are immobilized on nitrocellulose filters by a process called Southern blotting, named after its inventor, Edwin Southern. The filter is then exposed to a radioactively labelled gene probe and the position of the fragments analysed by placing the filter under an X-ray plate (Figure 15). By using a series of different enzymes that cleave DNA either within or outside a gene or genes, and orientating the fragments with respect to each other, it became possible to build up what were called restriction-enzyme maps of areas of the genome.

Two further developments in the 1970s were critical for the future analysis of the defective genes in the patients with thalassaemia. First, techniques became available for the rapid sequencing of DNA. But, of course, if the fine structure of normal and abnormal genes was to be studied, it was necessary first to isolate them. The second major advance, therefore, was what became known as gene-cloning. In essence, this involved inserting foreign DNA—that is, the DNA containing the gene to be isolated—into an appropriate vector, a DNA molecule capable of replicating in a bacterial cell. Replication of the inserted fragment along with the vector in the bacterium made it possible to amplify and isolate the foreign DNA.

Over subsequent years all these techniques were refined and expanded and provided the tools for the further exploration of the structure of genes, both normal and abnormal.

Not surprisingly, at the time they were developed, many of these new analytical methods were both complex and time-consuming. Gene-mapping might take a week or more, and cloning and sequencing a single human gene, depending on its size, considerably longer. But in the 1980s a technique was

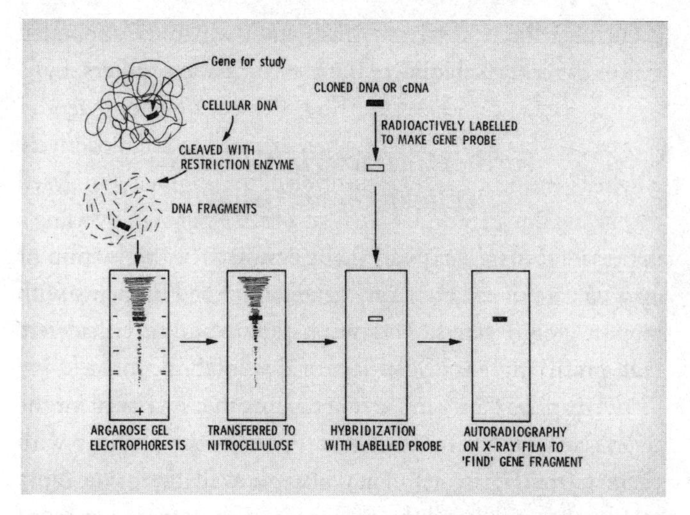

15. The principles of restriction-enzyme mapping. DNA for study is cleaved with a particular enzyme, and the fragments separated by their size on agarose gels. The latter pattern is transferred onto nitrocellulose, and the particular gene or gene fragments are searched for by hybridization with a radioactively labelled gene probe. The position of the radioactive signals is identified by placing the nitrocellulose onto an X-ray film. In this way maps of individual genes or parts of them can be rapidly constructed.

developed called the polymerase chain reaction (PCR), which allowed the amplification of short DNA sequences to be achieved over a period of just a few hours. Such was its power that it made it possible to amplify sufficient DNA from one or two cells to obtain a genetic diagnosis within twenty-four hours. Again, this technique has been modified and greatly improved since the 1980s.

This remarkable period in the evolution of molecular biology, in which many of the major technical developments led to the award of Nobel Prizes to their discoverers, set the scene for the major advances in gene discovery that were to occur over the next twenty years and that would ultimately lead to the remarkable era of the human genome project.

Further Characterization of the Human Globin Genes

In the late 1970s it was possible, by Southern blotting, to build up a picture of the physical organization and linkage of the human globin genes. This work provided direct evidence that, gratifyingly to those who had worked in this field for many years, all the gene arrangements that had been tentatively surmised from earlier genetic studies were more or less correct (see Figure 13). They also showed that, like most mammalian genes, all the globin genes contain one or more non-coding sequences, or introns (mentioned in an earlier section).

Following the advent of gene-cloning and sequencing, by the beginning of the 1980s the complete nucleotide sequence of the five human non-α globin genes and parts of their flanking regions had been determined. These studies showed that the non-coding regions of these genes contain a number of regulatory regions that are homologous with those found in genes of other species. At the same time equally rapid progress was made in analysing the arrangement of genes in the α globin gene cluster and in determining the sequence of the linked α globin genes and of the DNA which surrounds them.

During the early mapping studies of the human globin genes, a discovery was made that, in the longer term, was to become of enormous value in studying the molecular pathology of the thalassaemias, in particular, and for the future discovery of genes for many other diseases. In 1979 Jeffreys found several polymorphisms for a particular restriction enzyme in the globin genes of a number of individuals and estimated that about up to 1 per cent of the nucleotide sites in humans might be polymorphic. The word 'polymorphism' in this context simply means that there is an apparently harmless nucleotide base change in a sequence of DNA, and, because this either provides a new site for attack by a restriction enzyme or removes a previously existing one, these variations in DNA sequence became known as restriction fragment length polymorphisms (RFLPs). It was subsequently found that such RFLPs are spread throughout the globin gene clusters, but that, surprisingly, they do not occur in a random fashion. Rather, they are present in linked groups called haplotypes, and, surrounding individual genes or gene families and within any particular population, there are usually a small number of haplotypes that are common and a much larger number that are rare. As well as providing a valuable approach to studying population affinities and evolution, as we shall see later, the linkage of particular haplotypes to different forms of thalassaemia provided a valuable way of studying thalassaemia mutations, both within individual families and in populations.

A related twist to this evolving story came in 1978, when Kan and Dozy, using a particular restriction enzyme, found that the β globin gene of normal persons is located on a fragment of DNA of approximately 7.6kb (kb = kilobase or 1,000 nucleotide bases). However, if the same enzyme was used for cutting the DNA of patients with sickle-cell disease, the β globin gene was on a larger piece of DNA of approximately 13kb. It appeared, therefore, that the sickle mutation was on a chromosome that also carried a polymorphism of the DNA at a short distance from the coding area for the β globin chain. In the event, this was an absolutely seminal finding that opened up a completely new era in the field of human genetics. It was soon appreciated that, if there were polymorphisms closely linked to a gene for a particular disease, it might be possible to use this type of linkage approach to search for the gene or genes for any genetic disease.

Mapping the Globin Genes in Thalassaemia

Gene-mapping technology started to be applied to the study of thalassaemia in the late 1970s and rapidly confirmed and extended some of the observations that had been obtained a year or two earlier by hybridization technology. This was particularly true in the case of the α thalassaemias, because many of the common forms turned out to be due to gene deletions. It was found that, in the severe forms of α thalassaemia, α^0 thalassaemia, the α gene sequences were completely deleted in the homozygous state. It was subsequently found that one of the pair of genes is deleted in the milder form of α thalassaemia, α^+ thalassaemia. Then it was discovered that the latter condition is heterogeneous and that it results from two different-sized deletions of the α globin genes. Soon, some of the long-standing puzzles about the nature of α thalassaemia in African populations were solved. It was found that those who have the clinical picture of apparent heterozygosity for α^0 thalassaemia are, in fact, homozygotes for the deletional forms of α^+ thalassaemia. This condition turned out to be extremely common in African populations, in whom the absence of the α^0 mutation explained the lack of Hb H disease; it was now clear that this disorder results from the inheritance of an α^{0} thalassaemia deletion from one parent and an α^{+} thalassaemia deletion from the other.

As these gene-mapping studies were applied to the analysis of α thalassaemia from patients from other parts of the world, it became clear that there are many different-sized deletions that can cause α^{0} thalassaemia and that α^{+} thalassaemia is equally heterogeneous.

Studies of the β thalassaemias by gene-mapping were not so productive, largely because very few of them turned out to be due to gene deletions. However, as already mentioned, our rather crude hybridization studies carried out earlier in the 1970s had suggested that there might be one form of β thalassaemia that results from a defect at one end of the β globin gene. It was reasoned that this might be a good candidate for a partial deletion of the gene, and restriction-enzyme mapping showed that this was indeed the case; approximately 600 nucleotide bases had been lost from this region. Subsequent work suggested that this particular mutation is common in people from north India, but extensive mapping studies of other DNA samples obtained from patients from all over the world, at least at this stage, showed no cases that were due to deletions of the size that could be identified by this approach.

Mapping was also applied to a study of δβ thalassaemia and hereditary persistence of fetal haemoglobin and confirmed earlier hybridization studies that suggested that at least some of these conditions are the result of deletions that involve the β globin gene cluster. Indeed, by 1980 it was already apparent that these diseases, like the commoner forms of thalassaemia, are extremely heterogeneous at the molecular level.

Cloning and Sequencing the Globin Genes of Patients with Thalassaemia

It was already apparent by the late 1970s that the majority of the β thalassaemias were unlikely to be due to gene deletions and therefore it would be essential to be able to sequence the β globin genes involved to provide an answer to the underlying mechanism of the disease. The complete sequence of the different globin genes and their flanking sequences were published by several groups in 1980. The first successful identification of a mutation causing β^+ thalassaemia by gene sequencing was reported by Spritz and his colleagues and by Westaway and Williamson in 1981. Both groups, working independently on samples from Cypriot patients, identified a single nucleotide base substitution, G>A, at nucleotide 110 of the first intervening sequence of the β globin gene. At first this was a very puzzling result, because, as we have seen, the intervening sequences, or introns, that are found in all mammalian genes do not code for amino acids, and their function was at that time, and to a lesser degree still is, unknown. However, it was already known that, when mRNA is transcribed from its DNA template, the initial transcript contains intron sequences that have to be removed in the nucleus and the coding sequences, or exons, stitched together to form the definitive template for protein synthesis in the cytoplasm of a cell (see Figure 13). Could the new intron sequence that had been discovered somehow lead to abnormal splicing of mRNA during the process whereby the introns are removed and the exons joined together? Remarkably, this hypothesis was proved to be correct within the same year as the mutation was discovered.

This was a very encouraging start, but before long a major problem arose. Cloning and sequencing of globin genes from patients with thalassaemia were extremely time-consuming. Carrying out this procedure on random individuals from similar racial groups would be likely to lead to the repeated identification of genes with the same mutation, particularly if certain forms of thalassaemia are prevalent in the same population. This concern was soon found to be more than justified; within a period of only one year, six different research groups found the same mutation in the β globin gene from unrelated individuals of Mediterranean background!

This problem was solved by Orkin, Kazazian, Antonarakis, and others only a year or two later. As mentioned earlier, it had been observed as soon as extensive gene-mapping studies of the globin genes of normal individuals were carried out that the polymorphic restriction enzyme sites in the globin gene cluster are not arranged in a random fashion but in a series of patterns, or haplotypes, that vary in frequency and constitution between different populations. It was reasoned that the different types of β thalassaemia mutations would be likely to be found in association with particular haplotypes. Furthermore, the haplotypes could be used within families to mark individual β globin genes and hence provide a way of sequencing the appropriate gene. In the event, these predictions turned out to be correct; for example, it was found that in Mediterranean populations different β thalassaemia mutations were associated with a limited number of different haplotypes. With this approach, the discovery of β thalassaemia mutations moved extremely quickly, and by the end of the 1980s almost 100 different ones had been discovered; the number is now over 200. It was already becoming clear that, in

TABLE 3 Some examples of the different classes of mutations that cause the thalassaemias

Gene deletions	Loss of all or part of a globin gene.
Nonsense mutations	Base substitutions that lead to premature cessation of synthesis of a globin chain.
Mis-sense mutations	Base substitutions that result in a different amino acid being incorporated into a growing α or β chain, sometimes leading to its instability.
Frameshift mutations	Base substitution or deletion that alters the reading frame of the genetic code and prevents α or β chain production.
Splice junction mutations	Base substitutions in exons or introns that interfere with the processing of messenger RNA.
Regulatory mutations	Base substitutions or small deletions that interfere with the normal transcription of messenger RNA for an α or β globin chain.

any population in which there is a high frequency of β thalassaemia, one or two mutations predominate.

What kinds of mutations were found in the β thalassaemias? It appeared that anything that could possibly go wrong with a gene might be found if enough patients with the disease were studied (Table 3). In some cases a single DNA base change resulted in premature termination of globin chain synthesis or scrambling of the genetic code that resulted in a defect in its reading frame. A wide variety of mutations was also found in the critical coding regions at the junctions between introns and exons, which are involved in excision of introns and joining of exons to form the definitive mRNA template for globin-chain synthesis. Furthermore, it soon became apparent that some forms of thalassaemia can result from mutations outside the globin genes themselves in regulatory regions that are involved in the control of their activity and that, in effect, reduce the rate of globin chain synthesis. Yet others could result in the production of highly unstable β globin chains that precipitate or are rapidly destroyed in red-cell precursors.

In the early 1980s it was found that, although the common forms of α thalassaemia result from gene deletions, in some forms the genes are intact, yet there is a variable reduction in the output of α globin chains. When these potentially defective α globin genes were cloned and sequenced, it was found that they contained mutations very similar to those that had been found in the β thalassaemias. Furthermore, studies of this kind confirmed the suggestion that there is a class of α thalassaemias that are due to mutations in the chain termination codon, a suggestion that had been made previously by studies of the protein structure of Hb Constant Spring, as described earlier in this chapter. Similar progress was made towards a better understanding of those forms of hereditary persistence of fetal haemoglobin in which no deletions of the β globin gene cluster could be demonstrated. In this case, mutations were found in the regulatory regions of the γ globin genes, which, presumably, were responsible for their persistent activity after the neonatal period.

Over the next ten or fifteen years laboratories from all over the world developed the technology to analyse globin genes by these new techniques, and an extraordinary picture of the diversity of mutations gradually emerged. Furthermore, it became apparent that the original concept of there being only one or two common mutations in each high-frequency population was correct. We will return to consider the evolutionary implications of this remarkable finding in a later chapter.

A Better Understanding of the Clinical Diversity of the Thalassaemias

As we saw in earlier chapters, from the first descriptions of the thalassaemias, particularly in Italy, it was apparent that they reflect a spectrum of disease ranging from profound anaemia in the first year or two of life and early death, through more moderate forms of anaemia that are compatible with survival into adult life, to the extremely mild forms of the condition that were later found to represent the heterozygous carrier states. The terms 'thalassaemia major', 'intermedia', and 'minor', which reflect this clinical heterogeneity, though purely descriptive, are still used today to describe these different clinical forms of the disease.

A particularly good example of the clinical diversity of what seems to be the same form of thalassaemia was first described in 1956 in early studies in Thailand. In the Thai population, and as discovered later, in populations stretching from the eastern side of the Indian subcontinent through Bangladesh, Burma, and Thailand, and down the Malay Peninsular into Indonesia, the commonest form of severe thalassaemia is due to the inheritance of a β thalassaemia gene from one parent and a gene for the structural haemoglobin variant, Hb E, from the other. Even from the very first description of this disease in Thailand, it was clear that, although these patients seemed to have received identical genes from both parents, there is a very wide diversity of their clinical pictures, ranging from profound anaemia in early life to a disorder that is compatible with reasonable growth and development. Although there has been some progress towards a better understanding of the reasons for this clinical diversity, it is a problem that still faces those who care for patients with this condition in Asia today. And this remarkable clinical diversity is not restricted to β thalassaemia and its variants. The more severe form of α thalassaemia, Hb H disease, which also occurs widely in South East Asia, also shows remarkable clinical diversity.

Since the profound anaemia of β thalassaemia results from imbalanced globin chain synthesis and the damaging effects of the α globin chains that are produced in excess, one obvious mechanism that might underlie its clinical diversity would be a variability in the degree of defective β globin chain synthesis as the result of the underlying mutation of the β globin genes. When the mutations of these genes were characterized after the 1980s, it was found that those that were associated with only a mild reduction in β globin synthesis were mainly in the regulatory regions of the β globin genes, although a few were found in the intron/exon junctions or close to them but appeared to cause only minor defects in RNA splicing. However, only a limited number of these 'mild' mutations were discovered, and many patients were encountered with mutations that caused either a total or a severe reduction in β chain production yet were associated with a milder clinical picture. How could this be?

As early as 1961 Fessas described a mildly affected patient with β thalassaemia whom he suspected had also inherited α thalassaemia. Since the severity of β thalassaemia reflects the degree of imbalanced globin chain synthesis, it was suggested that, perhaps, some milder forms of β thalassaemia might result from the co-inheritance of one or other form of α thalassaemia. Despite the difficulties in defining the α thalassaemias during the 1960s and 1970s, numerous family studies were reported over this period that suggested that at least some forms of relatively mild β thalassaemia might reflect interactions of this type, observations that were put on a firm basis in the early 1980s, when it was possible to analyse the α globin genes directly. Of course, the degree of amelioration of β thalassaemia by α thalassaemia depends, at least to some degree, on the severity of the β thalassaemia mutation; patients who are homozygous for a β^{0} thalassaemia gain little advantage, with the possible exception of presenting later with the disease, while those with relatively less severe forms of β^{+} thalassaemia may have a much less severe clinical disorder if they have co-inherited α thalassaemia.

In the 1960s and 1970s there were also reports of patients who appeared to be homozygous for β^{0} thalassaemia who, nevertheless, had a relatively mild disease, which seemed to be associated with the ability to produce unusually high levels of fetal haemoglobin. It became clear that an infant who was able to produce more γ chains of Hb F than usual after birth would be protected to some degree from the effects of severe β thalassaemia; the γ chains would combine with α chains to produce Hb F, so reducing the degree of globin chain imbalance. But, although polymorphisms in the regulatory regions of the γ globin genes that had some effect in modifying the level of Hb F in β thalassaemia were discovered, the effect was small; the reasons for the variation in Hb F levels remained elusive.

It was also realized in the 1960s that, if it were possible to find a way of maintaining γ chain synthesis early in life, or switching it on again, this would provide an extremely valuable therapeutic approach to the treatment of β thalassaemia. One of the major problems in achieving any form of control over the regulation of fetal haemoglobin production was, and still is, lack of a genuine understanding of what regulates the switch from fetal to adult haemoglobin in normal infants. Indeed, this is probably one of the most disappointing aspects of the haemoglobin field; intensive studies over close on half a century have still not provided a picture of what throws this switch and how it is regulated. It was long believed that a detailed study of hereditary persistence of fetal haemoglobin and its underlying causes might provide some inkling about the genetic control of fetal haemoglobin switching. But, although many different forms of this condition were studied and their underlying molecular pathology defined, it has not been possible to explain why they are associated with elevated levels of fetal haemoglobin production.

But one thing became absolutely certain after all this effort; an increased ability to produce fetal haemoglobin in adult life is undoubtedly a major factor in modifying the clinical picture of severe β thalassaemia for the better. Even at the time of writing this book, although it is now apparent that many different genes must play a role, the way in which they act to elevate fetal haemoglobin in adult life is still not clear.

Although by the early 1980s it was established that the clinical picture of β thalassaemia can be modified by coexistent α thalassaemia or an increased ability to produce fetal haemoglobin, this was by no means the end of the story of the clinical diversity of the different forms of thalassaemia. As early as 1973 it was recognized that some relatively severe forms of β thalassaemia are inherited in a dominant fashion—that is, only one defective gene is required to produce the disease. It was later found that these varieties of thalassaemia result from the production of elongated and highly unstable β chains, which precipitate in the bone marrow together with excess α chains. Later, when it became possible to study the α globin genes by DNA analysis, it was found that other patients with β thalassaemia who have only a single defective β gene may have a variably severe clinical picture because they have inherited extra α genes; most of us have four, but in some populations there is a variable frequency of individuals who may have five, six, or more.

From the 1980s onwards it also became apparent that some of the most severe complications of the β thalassaemias might be modified by genetic factors that acted through completely different mechanisms than those involving globin synthesis. For example, evidence started to appear that genetic factors might play a role in determining the relative rates of iron absorption from the bowel, modifying the severity of bone disease, and in determining the severity of the level of jaundice because of the excessive destruction of red cells and their precursors that occurs in many forms of thalassaemia of intermediate severity. In short, the clinical picture of a child with severe α or β thalassaemia is the result of a primary defect of one or other of the globin chain genes set in the background of layer upon layer of complex interplay between many other aspects of their genetic makeup. Curiously, over this period less attention was paid to the environment or about how children with these diseases adapt to their anaemia and other manifestations of the disease; such research was only to come much later.

This gradual unravelling of some of the factors that are responsible for the remarkable clinical diversity of a single gene disorder like β thalassaemia was extremely important at the time. In effect, it showed that even a so-called simple monogenic disease is in fact far from simple and that the clinical picture may be modified by many other genes as well as the environment.

Abnormal Gene Action and the Pathophysiology of Thalassaemia

While the advances in our understanding of the molecular basis of the thalassaemias during the late 1970s and 1980s were particularly spectacular, it was equally important to try to understand how the imbalanced globin chain production that characterizes every form of the disease results in the characteristic clinical manifestations of these conditions. As mentioned in an earlier chapter, during the 1960s there was some progress made towards an understanding of the basic principles of the ineffective red-cell production and shortened red-cell survival that underlies the profound anaemia of the thalassaemias. During the 1970s and 1980s there were major advances in our understanding of the fundamentals of iron metabolism, the structure and function of the red-blood-cell membrane, the mechanisms whereby cells degrade abnormal proteins, and many other aspects of red-cell physiology that had important implications for the thalassaemia field. Over this period teams led by Rachmilewitz and Hershko in Israel, Schrier and Yuan in the USA, Wickramasinghe and Porter in the UK, and many others utilized and extended this new knowledge towards a better understanding of the basic pathophysiology of the thalassaemias.

It is beyond our scope to describe the large literature that evolved from these studies from the 1970s onwards. These studies provided a detailed picture of the mechanisms of how excessive α or β chains result in the destruction of red-cell precursors in the bone marrow and the shortened survival of the products of these cells in the peripheral blood. They emphasized the oxidant properties of the degradation products of free α chains and the mechanisms whereby excessive iron leads to damage to the liver, endocrine glands, and heart. And they went some way towards the beginnings of an understanding of the reasons for excessive iron absorption associated with proliferation of the bone marrow. Over this period there was also a detailed analysis of endocrine function carried out in children with thalassaemia that provided considerable insight into the abnormalities of growth that occur in iron-loaded patients owing to damage to the pituitary and thyroid glands.

While these advances in our understanding of the pathophysiology of thalassaemia offered a much better understanding of its associated clinical features, they still left a number of intriguing questions. For example, the precise mechanism whereby a widely expanded bone marrow leads to the characteristic deformities of the face and skull still remained a mystery. And the remarkable pigmentation of the skin, which was emphasized by Cooley in his first descriptions of the disease, turned out to be related to increased melanin deposition rather than to iron itself. However, it had been noted since the mid-1970s that the skin rapidly lightens after treatment with chelating agents that remove iron, and it was suggested that the increased melanin deposition results from free-radical, or oxidative, damage similar to that responsible for sunlight-induced melanin synthesis; this reaction may be catalysed in thalassaemic children by a slight increase of iron in their skin.

By the turn of the century, therefore, nearly all the clinical features of severe thalassaemia that had been so carefully described by Cooley and his successors could be ascribed to the effects of globin chain imbalance and excessive iron deposition in the organs. While many details of these processes still remained to be worked out, there was now a reasonable understanding of how mutations of the globin genes could lead to the extraordinary diversity of the clinical manifestations of the thalassaemias.

While these successes in the thalassaemia field in the period between 1970 and the end of the twentieth century undoubtedly played a major role in the evolution of what became known as molecular medicine, historians of the period will undoubtedly wish to explore the influence of this new direction of medical research on improvements in the control and management of the disease. This is the subject of the next chapter.

VII

THE CONTROL AND MANAGEMENT OF THALASSAEMIA IN THE CELLULAR AND MOLECULAR ERA

Our story so far has focused on some of the remarkable developments in clinical research, protein chemistry, and molecular and cell biology, which laid the foundation for the gradual elucidation of the molecular and pathophysiological basis for the different forms of thalassaemia. But the second half of the twentieth century was also a period of rapid developments in other basic biological sciences that were to play an important role in the story of thalassaemia—notably, cell biology and immunology. In this chapter we will explore in outline the diverse routes that led to the first reported cure of a patient with thalassaemia in 1982. And we will also consider how the developments in the molecular analysis of the thalassaemias and a better understanding of their pathophysiology led to further improvements in their control and management.

Thalassaemia is Cured for the First Time: Bone-Marrow Transplantation

The research that led to the first successful human bone-marrow transplantations in the 1970s encompasses the whole field of the development of haematology, and later immunology, and can be traced back to the second half of the nineteenth century. The work of the great morphologists, those who studied the characteristics of the different cells of the blood, was dominated in the second half of the nineteenth century by the work of Paul Ehrlich, who believed that, based on their appearances, the various cells of the blood arise from different and distinct progenitors. Early in the twentieth century, however, other morphologists were attracted to what was called the monophyletic theory of blood-cell production—that is, that all the formed elements of the blood are the progeny of a single variety of progenitor cell. For forty years there was a bitter schism between those who held these different views. The argument ran largely unabated until the late 1950s, when it was found that animals can be protected against otherwise lethal doses of irradiation by grafting bone-marrow cells into irradiated recipients. It was later found that the latter developed nodules in their spleens, and that each nodule, or colony, originated from a single cell. By using cells with marked chromosomes from donor animals, it was confirmed that the same stem cell, termed a colony-forming unit in spleen, or CFU-S, could give rise to all the different cells of the blood.

Also during the mid-1950s, based mainly on the work of Jean Dausset in Paris and of Johannes van Rood in Leiden and many others, the human leukocyte antigen (HLA) system was discovered, work that was to open up the whole field of human organ transplantation by making it possible to identify the likelihood of compatibility between donors and recipients.

It is not surprising, therefore, that at about this time thoughts turned to the possibility of engrafting human bone marrow for the treatment of haematological disorders. Although many haematologists became interested in this possibility, it was E. Donnall (Don) Thomas (1920–) who must take the major credit for this remarkable advance in the treatment of blood diseases. Thomas trained at Harvard Medical School and, after serving in the Second World War and then undertaking further postgraduate training in haematology, in 1955 he was appointed to the Mary Imogene Bassett Hospital in New York as Physician-in-Chief. It was there that he started his transplantation experiments, at first in mice and later in outbred dogs. After eight years of work on these models, Thomas moved to Seattle, where he set up a major centre for bone-marrow transplantation research. Over subsequent years his team examined every aspect of human transplantation, and, after innumerable setbacks, they had their first successes in treating patients with aplastic anaemia and later with leukaemia. These early breakthroughs depended on the development of better ways of suppressing the recipient's bone marrow, the application of HLA-typing, and, not least, the meticulous aftercare of patients whose marrows had been suppressed during the transplantation and who were therefore at very high risk for infection. Following the pioneering work of Thomas and his many associates, centres for marrow transplantation were later established in the USA and in many other countries throughout the world. This work was later recognized by the award of the Nobel Prize for Medicine to Thomas.

The first successful bone-marrow transplant for thalassaemia was reported by Thomas and his colleagues in 1982. The patient was a 14-month-old child who had never received a blood transfusion; the results were remarkably successful. On the other hand, at about the same time a 14-year-old child with thalassaemia who had already received 150 red-cell transfusions received a bone-marrow transplantation in Pesaro, Italy, but this was followed by recurrence of the patient's thalassaemia after the bone-marrow graft had been rejected. As well as rejection of the graft by the patient's own immune system, a number of other complications dogged early attempts at marrow transplantation. In particular, a condition called graft-versus-host disease occurred quite commonly, which was due to an attack launched by the immune cells in the graft on its recipient. Furthermore, patients were particularly prone to infection after this procedure, partly because they had to receive marrow-suppressant drugs to prepare for the graft, but also drugs to suppress the immune response of both the recipient and the transplanted cells of the donor.

After the pioneering studies in Seattle, the main centre of research in bone-marrow transplantation for thalassaemia moved to Italy, notably Pesaro, under the direction of Guiddo Lucarelli. In the decade following these early experiences, over 1,000 transplants were performed at three centres in Italy; a few centres in other parts of the world gained experience, though of much smaller numbers. In reviewing the reasons for the steady improvement in the results of bone-marrow transplantation for thalassaemia over this period, Giardini suggested that, as well as increasing experience, the major factors had been the establishment of more effective pre-transplantation regimens, the introduction of more potent and effective immunosuppressant drugs, and improved treatment of infection. Whatever the reasons, the procedure gradually improved, and, by knowledge gained about the ideal type of patient for engraftment and the best time for the transplantation, in the best centres the procedure yielded over 80 per cent success rates.

Like most advances in modern medicine, the introduction of bone-marrow transplantation for thalassaemia was not without its controversies. Although the issues were complex, the central ethical problem is easily stated. Was it right to subject a baby or a young child to a procedure that still had a small but significant mortality rate, and a not inconsiderable morbidity, when, in countries that were rich enough to be able to offer this treatment, many of these patients could be maintained indefinitely in good health by adequate transfusion and chelation therapy? Also, could we be confident that there would be no long-term risk of children developing secondary malignant disease of the bone marrow after treatment of this kind? The proponents of marrow transplantation felt that, provided the transplant was successful, the dramatic improvement in the quality of life that followed for a child with thalassaemia outweighed these potential risks. As the procedure has gradually been improved, and mortality and morbidity have declined over recent years, they were undoubtedly justified in continuing their programmes.

The major problem, of course, was that children are suitable for a bone-marrow transplantation only if there is a donor of the appropriate HLA type. Since this will occur on average in only one out of four of their siblings, and the results of transplantation in non-matched donors are very poor, bone-marrow transplantation remained an option for only a limited number of children with thalassaemia.

Improvements in the Population Control of Thalassaemia

As we saw in a previous chapter, during the 1970s there was considerable interest in the development of approaches for the reduction in the number of births of babies with serious forms of thalassaemia. An approach that offered population screening and marital advice did not seem to have had a great deal of effect on the numbers of births of babies with this condition. On the other hand, parental screening with the offer of prenatal diagnosis of affected fetuses had been applied successfully, with a major reduction in the number of births of babies with this condition, particularly in Cyprus and Sardinia and to a lesser degree in other populations.

In the second half of the 1970s, after the first successes in identifying thalassaemia mutations at the DNA level, thoughts turned to the application of this new technology to the prenatal diagnosis of thalassaemia. The first diagnoses were carried out by using amniocentesis—obtaining a sample of the amniotic fluid that surrounds the fetus through a needle passed through the mother's abdominal wall. This procedure offered a safer option than fetal blood sampling, because the fetal loss rate is only about 0.5 per cent above the spontaneous mid-trimester figure. At about 16–20 weeks gestation, a 20-ml amniotic fluid sample contains sufficient fetal cells to provide enough DNA for analysis; an aliquot could also be cultured to provide additional DNA within two weeks, if needed. The first group of globin-gene mutations to be characterized at the molecular level in this way were the major gene deletions, using the technique of molecular hybridization as described in the previous chapter. In 1976 Kan and his colleagues studied a fetus at risk for the severe form of α thalassaemia, α° thalassaemia, and were able to show that it had the α thalassaemia trait. Gene-mapping by the Southern technique was soon applied for prenatal diagnosis, and fetuses with α° thalassaemia and δβ thalassaemia were diagnosed successfully by restriction endonuclease analysis of amniotic fluid cell DNA.

As mentioned in the previous chapter, in the mid-1970s Kan and Dozy had discovered that the sickle-cell gene exists on a chromosome with a linked DNA marker, and this observation was soon applied for the diagnosis of sickle-cell disease in fetal life. During this period more polymorphisms were discovered of this kind, linked to different forms of β thalassaemia, and hence this approach started to be applied more generally for the prenatal diagnosis of different forms of thalassaemia. Remarkably, by 1982 175 cases of amniocyte DNA diagnosis had been reported, without any diagnostic errors, confirming the reliability of linkage analysis for the diagnosis of genetic disorders.

Although these reports were very encouraging, the problem remained that, like fetal blood sampling, amniocentesis could not be carried out until the middle trimester of pregnancy. Hence, if parents decided to terminate an affected fetus, the procedure was often very distressing for the mother. It became important, therefore, to try to develop a method of obtaining fetal tissue earlier in pregnancy, between 8 and 12 weeks gestation. The aspiration of chorionic villi for this purpose provided the next major advance in the field.

The chorion, which surrounds the fetus during its early development, is composed of an outer layer, or trophoblast, an inner or mesodermal layer, and fetal blood vessels that extend all over the gestation sac until the end of the second month of

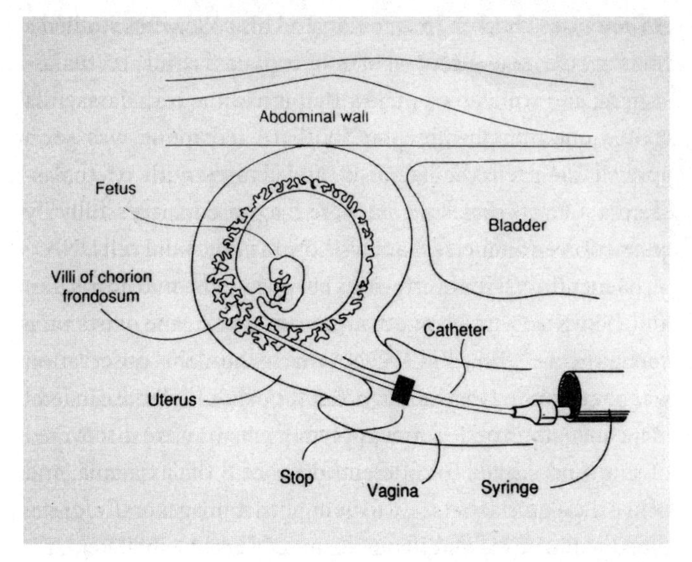

16. Chorionic villus sampling. This illustrates how a small sample of villi are obtained for DNA analysis for prenatal detection of genetic disease.

pregnancy. Since these tissues are all of fetal origin, if they could be biopsied, they would offer an invaluable source of fetal DNA early in pregnancy. It turned out that chorionic villus sampling (CVS) could be carried out by either transcervical or transabdominal routes, both under ultrasound guidance (Figure 16). Most of the early experience with this technique was gained by the transcervical approach, but the later development of transabdominal sampling had the advantage that it could be used up to 14 weeks gestation.

Naturally there were concerns about the safety, for both mother and fetus, of this new invasive procedure. Because there is a significant spontaneous fetal loss in the first trimester, a number of trials were needed to determine whether CVS would be associated with a greater loss than amniocentesis. A major study in Canada, published in 1989, indicated that there was no difference in maternal morbidity between patients who had undergone amniocentesis or CVS, and that there was no significant difference in the rate of fetal loss between the two procedures. A few years later there was some concern about the possibility of an increased frequency of limb-reduction defects in a fetus if CVS was performed between 55 and 66 days gestation. Since there was some evidence that these limb deformities tended to occur when a CVS was done in the ninth week of pregnancy or before, it was advised that the procedure should not be done until at least 10–11 weeks gestation, to minimize this risk.

The first successful prenatal diagnoses of β thalassaemia and sickle-cell anaemia following DNA diagnosis of tissue obtained by CVS were reported by Old and his colleagues in Oxford in 1982. This approach was soon taken up by other centres, and in 1986 the Oxford group and their colleagues published their experience of over 200 first-trimester diagnoses for haemoglobin disorders, including 133 cases at risk for β thalassaemia and 55 for sickle-cell anaemia. A successful diagnosis was made in all but one case; there were two failures, because of insufficient material, but a diagnosis was made subsequently by amniocentesis.

The development of first-trimester prenatal diagnosis for thalassaemia was undoubtedly a major step forward. It was much more acceptable to women than methods that could be used only later in pregnancy; they had not had so long to adapt to the fact that they were carrying a living being, and, if termination was required, it was very much easier. Hence, CVS became the method of choice for prenatal diagnosis, with amniocentesis used as a backup for women who presented too late for this procedure. Furthermore, during the 1990s and after, there were steady improvements in the technology of DNA analysis, which had the added advantage of improving both the speed and accuracy of prenatal diagnosis using this approach.

Further Ethical and Religious Issues Relating to Screening and Prenatal Diagnosis

As mentioned earlier, the first attempts at population screening and counselling in an attempt to reduce the number of marriages between carriers for thalassaemia or sickle-cell disease appeared to have little effect on the number of births of affected babies. With the exception of Iran, there were no more major efforts towards this end.

In 1997 Ghanei and his colleagues reported on their experiences of population screening in the Isfahan province of Iran, a mainly Islamic country. Over 100,000 individuals preparing for marriage were screened for β thalassaemia trait, and at-risk couples were referred for further consultation. After three years the proportion of high-risk couples initially deciding not to marry was 90 per cent, and no new cases of thalassaemia were detected in the children in the populations that were screened. This remarkable outcome raises some important issues, however. For example, the authors wrote that 'those couples in which both members had the trait who wanted to marry would have to register their marriage and perform the religious ceremony in another area outside the district of Isfahan'. Apparently this was because the marriage officers had agreed not to register such couples because of the high risk of a birth of a child with thalassaemia major, the prohibition of abortion, and the risks associated with illegal abortion.

In discussing their results, the authors suggest that the high rate of those choosing not to marry, 90 per cent compared with <10 per cent in Cyprus, may have been related to the Iranian custom of there being no long-term relationships between couples before marriage. While this may be the case, it is not clear from this paper how much pressure was brought to bear on these couples not to marry; the fact that they could not marry in their own province if they were both carriers must surely have had a major effect on their decision. Interestingly, and as we shall discuss later, there was a major change in thinking in Iran and other Islamic countries: prenatal diagnosis became acceptable only a few years after this work was reported.

But to what extent is it acceptable to avoid genetic disease by prenatal testing and termination of affected pregnancies? This issue was widely debated in Western society throughout the second half of the twentieth century. The extreme view was that, under no circumstances, should parents be allowed to decide to avoid or terminate a pregnancy in which a baby has a serious genetic handicap. Rather, society should care for these children, many of whom, it was said, can lead happy and fulfilled lives if properly looked after. These views were not confined to those who have had no experience of dealing with sick children, and were common among patients, doctors, and thoughtful members of society, regardless of whether they were based on religious or humanistic beliefs. Some of those who believed that termination of pregnancy for genetic disability is wrong took an extreme position. Surely, they argued, physical disability is not always a bad thing; some of our greatest creative artists suffered in this way. Is it right to terminate a pregnancy and risk losing a Beethoven? In a thoughtful essay Appleyard emphasized how irrational these arguments can become. He quotes from G. K. Chesterton, an author who apparently had an intense fear of the eugenics movement: 'Keats died young; but he had more pleasure in a minute than eugenicists get in a month.' Appleyard pointed out that a genius with tuberculosis is a phenomenon we like to feel is beyond any rational balancing of possibilities, and that it is natural therefore to imagine that an aborted fetus or a pregnancy prevented by genetic counselling might have become a Keats or an Einstein. But, despite these concerns, prenatal testing and termination of pregnancy became widely accepted in Western societies in the latter half of the twentieth century.

Of course, religious beliefs have always been a major issue in shaping programmes for the control of the thalassaemias. Of the various branches of Christianity, the Roman Catholic Church's position is absolutely clear: there should be no interference with a pregnancy, whatever the reason. The question of population screening and premarital advice was discussed with Pope Pius XII by participants at the 7th Congress of the International Haematology Society in 1958. He was not averse to screening and education provided no pressure was applied to families regarding reproduction and, in particular, no advice on contraception was included in the programme. But there are no signs that the Catholic Church will modify its views on termination of pregnancy; the same applies to the fundamentalists in the USA.

The attitudes of different branches of Christianity to this question are well exemplified in Cyprus and Sardinia. When screening and prenatal diagnosis were being established in Cyprus in the 1970s, the Orthodox Church contributed greatly to their success by encouraging premarital screening and developing educational programmes in the school curricula.

In contrast, in Sardinia the Catholic Church assumed what was described as 'an uninterested attitude' towards a similar campaign. Presumably much the same must have happened in parts of Italy. For example, new cases of β thalassaemia became extremely unusual in Ferrara, which was almost certainly due to a control programme instituted in the 1970s. Since the Catholic Church did not change its views, presumably this state of affairs reflects a laissez-faire attitude by its local priests combined with the emergence of a more secular society. This interpretation of events is compatible with the limited effectiveness of the thalassaemia control programme in Rovigo, where Catholic culture had a much stronger influence.

Although Islam does not formally permit any form of interference with natural events, it is particularly interesting that prenatal diagnosis for thalassaemia became acceptable in many Islamic countries in the 1990s. Apparently the Muslim jurisconsults of both major sects agreed that, if genetic testing proves definitely that a fetus is affected by a serious disease, abortion is permissible and lawful. More specifically, they stipulated that a pregnancy can be terminated only before the time of 'breathing the soul'—that is, before 120 days gestation. It appears, however, that the situation is not as straightforward as this in practice. Working in Cameroon, Salihu found that it is extremely difficult to convince a Muslim population with a high illiteracy rate that Islamic jurisconsults in Western capitals have found that it is permissible to condone the termination of pregnancy. This population tends to believe only what their local Imams tell them.

The position in Buddhist populations with high frequencies of thalassaemia is equally complex. The situation in Thailand and Sri Lanka, countries in which I have worked for many years, was discussed briefly by Wasi and Fucharoen in 1995. The form of Buddhism in these populations, Theravada, has the same root, and the Buddhist principle is to refrain from killing any living organism. Recently I had the opportunity of discussing this issue with the Dalai Lama, who confirmed this belief, but then added with a delightful smile that Buddhism is a very adaptable philosophy. In Thailand this certainly appears to be the case, because prenatal diagnosis has become acceptable provided that it is for the benefit of the mother and within the doctor's conscience. The situation in Sri Lanka is different, however, and, despite the very high frequency of illegal abortion, termination of pregnancy for medical indications is still not accepted. Hindus seem to have a more relaxed approach to these questions; it appears that both these religions are beginning to accept prenatal diagnosis under some circumstances.

Overall, therefore, screening, prenatal testing and termination of pregnancy for babies with severe forms of thalassaemia gradually became acceptable across most of the high-frequency regions of the world from the 1970s onward. There have been a few examples of governmental pressures on couples to undergo testing; in Cyprus this became a legal requirement in the 1970s for couples before marriage, but the finding that they were both carriers did not preclude them from marrying. They must, however, marry with the knowledge of their genetic make-up with respect to thalassaemia. Sadly, the major form of pressure on parents to undergo prenatal testing, at least in some populations where it is widely established, came in completely the opposite direction. In Sardinia, because it is the norm for couples to undergo this procedure, families who are unwilling to do so may come under considerable pressure and even be ostracized by other families in their village for allowing a child with a severe form of thalassaemia to be brought into the world.

As mentioned in an earlier chapter, the introduction of prenatal diagnosis using fetal blood analysis in the 1970s had a dramatic effect on the frequency of births of babies with β thalassaemia, particularly in Cyprus and Sardinia and to some extent in many Western countries. With the introduction of chorion villus sampling and improved methods of fetal DNA analysis, this trend continued, and there was a remarkable decline in the births of babies with thalassaemia throughout the Mediterranean region, and parts of the Middle East, as well as among the immigrant populations with thalassaemia in Europe and the United States. Similar progress followed, though much more slowly, in the Indian subcontinent and in countries in East and South East Asia. At least some of the impetus towards this approach to the control of thalassaemia was undoubtedly economic; a number of studies showed quite clearly that the establishment of screening and prenatal diagnosis were much more cost effective than the complex long-term management of thalassaemia. Furthermore, for many families the prospect of being able to have normal children was an even greater incentive towards the establishment of programmes of this type.

Progress in Symptomatic Treatment

While the dramatic expansion of knowledge about the molecular bases of the thalassaemias undoubtedly had a major influence on the more accurate genetic counselling of families with the thalassaemias, and greatly improved the facility for their prenatal diagnosis, this information led to few major improvements in the symptomatic management of the thalassaemias in the twenty-five years that followed an understanding of their molecular pathology. The same has to be said for the increasingly elegant studies that were carried out over this period that led to a detailed picture of the mechanisms of the destruction of the red-cell precursors and the shortened survival of the red cells in different forms of thalassaemia. There was, however, a genuine improvement in the symptomatic management of these diseases over this period, although the impetus came from directions other than molecular and cell biology.

As discussed in a previous chapter, once the most appropriate ways of administering regular blood transfusions to thalassaemic patients had been established, the major issue was how to avoid the complications of progressive iron loading. By the late 1970s the optimal way of administering desferrioxamine had been established, although, of course, iron loading occurs over a relatively long period, and it was not until the mid-1990s that the genuine value of this approach to iron chelation was fully established. In 1994 Garry Brittenham and his colleagues in the USA and Nancy Olivieri and her team in Toronto reported the results of the long-term administration of desferrioxamine by the subcutaneous route, and related survival to body iron stores measured either by liver iron levels or plasma ferritin estimations. These seminal studies showed that, provided the body iron was controlled at defined levels for hepatic iron or serum ferritin, there was close to 90 per cent survival, results that were confirmed in 1996 by similar studies in Italy. As mentioned earlier, drug toxicity had not been a major problem over this long period of study, and the major difficulties that had been encountered were lack of compliance because of the discomfort and complexities of the subcutaneous route for administration of the drug. It should be added that this relatively expensive treatment was available only in richer countries and, at least until 2005, in almost none of the developing countries.

These early successes of the use of desferrioxamine were a great stimulus to attempts to try to develop an iron-chelating agent that could be taken by mouth. Probably because they had no inkling of the picture of the global burden of thalassaemia that was evolving, and those that were aware knew that the bulk of patients lived in poor countries, apart from CIBA-GEIGY (as it was then called), no pharmaceutical companies showed any interest in the development of chelating agents. The first oral agent was developed by Hider and his colleagues in London in 1982. The drug that evolved from their work on hydroxypyridinones, deferiprone (L1), because it had not been developed by a pharmaceutical company, had a rather unconventional start to its life. It was difficult for university research workers to carry out anything more than very limited trials and animal studies without the backing of corporate support. Early trials showed promising results, as demonstrated by urinary iron excretion in patients with chronic neoplastic diseases of the bone marrow who had transfusional iron overload. Although a patent was finally taken up by CIBA-GEIGY, in 1993 the company decided not to develop the drug because of its findings in animal toxicity studies. Its further development was then taken over by another pharmaceutical company, and trials were begun in 1993 to establish its efficacy and potential toxicity compared with desferrioxamine. The study was discontinued in 1996 by the company when Nancy Olivieri, the clinical scientist involved, became worried about the lack of control of iron loading in a significant proportion of patients, together with concerns about increased fibrosis (scarring) of the liver. She was threatened with legal action if her results were reported to the scientific community, which they soon were. She was then dismissed from her post at the University of Toronto, and later reinstated, but only after outside intervention. A subsequent independent external review report completely vindicated her actions and, incidentally, disclosed the University's potential financial interests with the company. This is only one example of the increasingly uneasy relationships between academia and industry that were to affect thalassaemia research, and medical research in general, after the turn of the century.

Many smaller, uncontrolled clinical studies of deferiprone followed. Later, retrospective studies suggested that it might be more effective at removing iron from the heart than desferrioxamine, but long-term prospective studies of this important possibility were, for reasons that are not clear, never embarked on. This agent was also developed by companies in Asia and widely distributed, particularly on the Indian subcontinent. But its role was still not absolutely clear. Perhaps this is not surprising, given the time that it took to demonstrate the effectiveness of desferrioxamine. But it is clear that deferiprone is the more toxic of the two, and hence the determination of its role in the management of thalassaemia was bound to be more complex.

Although there were other attempts to develop orally active chelating agents, it was not until 2003 that the first reports appeared of a new oral chelator called Exjade, manufactured by Novartis. This agent would also have to go through many years of detailed study, in the same way as its predecessors.

Improvements in General Medical Care

It is very important to remember that, in addition to these more specialized forms of treatment, advances in the general medical management of children and young adults stemming from many other fields of medicine in the second half of the twentieth century played a vital role in improving the outlook for children with thalassaemia. In the early history of this disease, many children succumbed to infection, and the advent of powerful antibiotics and vaccines was a major factor in their longer survival. And, in many richer countries, the screening of blood for hepatitis and, later, HIV reduced the frequency of blood-borne infection. Many other developments were to play a role, including intensive care, new diagnostic technology, and the rapid evolution of expertise in fields such as cardiology, endocrinology, metabolic bone disease, and others that had important implications for managing the complications of different forms of thalassaemia.

Changes in the organization of haematology also played a role in providing better care for these children. Although this field started out as a laboratory specialty, certainly in the USA it evolved such that haematologists were trained both clinically and in the laboratory. And by the 1950s there was rapid progress in the evolving specialty of paediatric haematology. This period of change took much longer in countries like the United Kingdom, where there was a battle, often heated, as to whether the specialty should be laboratory- or clinically based, and it was not until the 1970s that haematologists became fully trained in clinical as well as laboratory practice. The evolution of clinical and paediatric haematology led to the development of specialized centres, at least in the richer countries, for the management of thalassaemia. Interestingly, in the developing countries the pattern of development of haematology also varied, depending on outside influences, including, in some cases, their colonial past. In Thailand, for example, the early influence of American training led to the development of clinical haematology in the 1950s, whereas, in countries where there was a stronger British influence, even today haematology remains a laboratory specialty.

These organizational changes were undoubtedly helped by the formation of parent support bodies for thalassaemic patients in many countries and by the development of organizations like Thalassaemia International Federation that arranged meetings all over the world for training doctors, nurses, and patient's families in different aspects of care.

Any patient with a chronic disease like thalassaemia requires continuity of care above all else. In some centres in the richer countries this was finally achieved by having clinical haematologists as the central carers for these patients, with the help of both paediatricians and adult physicians for specialized problems at particular stages of their development. Unfortunately, in many countries arrangements of this kind were slow to evolve, and, because of the tight demarcation between paediatrics and adult medicine, children who reached adolescence were often handed on to adult physicians with no knowledge or interest in the disease. This should not surprise us, however. In many of the richer countries patients with the common haemoglobin disorders were largely confined to the rapidly increasing immigrant populations. Their doctors had received no training or education in the recognition or management of these conditions. Indeed, although clinical genetics developed rapidly as a specialty in these countries after the Second World War, the level of medical education in this field was extremely slow to follow suit. Overall, however, and despite these difficulties, the control and management of the thalassaemias gradually improved towards the end of the twentieth century.

Although it has been possible to touch only on the ethical, social, and organizational aspects of the thalassaemia field in the second half of the twentieth century, there are several other influences that will undoubtedly be of interest to social historians of the future. Not the least, and as has been discussed in the case of chelating agents, is the increasing influence that the pharmaceutical industry began to yield in research and clinical practice over this period, a phenomenon that was certainly not restricted to the thalassaemia field. The overall effect was that research workers and practising doctors became increasingly dependent on industry for support, not just for clinical trials or research projects but for postgraduate education at every level. Similarly, both national and international parent associations made increasing use of support from the same source. This phenomenon reached its peak in many developing countries, where doctors became totally reliant on industry for any form of support, particularly for attending national or international meetings. The potential dangers of this situation began to be appreciated only in the new millennium. The effect that it had on the thalassaemia field in the later years of the twentieth century and after must be for historians of the future to determine, but undoubtedly this trend raises genuine difficulties in trying to assess the quality and genuine objectivity of at least some research and clinical practice in the field over this period.

Experimental Treatments

As mentioned in previous chapters, from 1960 onwards it became increasingly clear that patients with β thalassaemia or sickle-cell anaemia who were able to produce unusually large amounts of fetal haemoglobin almost invariably had milder forms of their disease. It is not surprising, therefore, that thoughts turned to whether it might be possible to stimulate Hb F production in one way or another. This notion was always difficult to put into practice, not least because of lack of any clear knowledge about the mechanisms of the switch from fetal to adult haemoglobin production. Hence, any progress that was made in this field had to rely on chance clinical or related observations that were made along the way.

In the mid-1970s it became clear that reactivation or modification of the level of fetal haemoglobin occurs under a variety of conditions, including recovery of the bone marrow after the use of cytotoxic agents for the treatment of leukaemia, following recovery after bone-marrow transplantation, and in animals after removal of a relatively large volume of their blood, a phenomenon particularly marked in baboons. Following these observations, research for the discovery of agents that might reactivate or elevate fetal haemoglobin synthesis followed side by side in the sickle-cell anaemia and thalassaemia fields.

It was reasoned that agents that cause perturbation of erythropoiesis, even if they have only an indirect effect on the expression of the γ globin genes, might have at least some effect on elevating Hb F production. Early studies with such an agent, 5-Azacytidine, showed some promise, but this particular drug was found to be too toxic for this purpose. Later it was found that the cytotoxic agent hydroxyurea had a remarkable effect in elevating fetal haemoglobin levels in baboons, particularly if they were bled at the time of administration. Based on these results, treatment with hydroxyurea of patients with sickle-cell anaemia was investigated, and some patients showed a promising elevation of fetal haemoglobin. These early results led to a carefully controlled clinical trial, the results of which, published between 1992 and 1996, showed quite unequivocally that this agent can play a major role in reducing the clinical severity of sickle-cell disease. It was not absolutely clear, however, whether this was entirely due to a modest elevation in Hb F production or whether the clinical efficacy reflected more subtle actions of the drug, including mild bone-marrow suppression and a reduction in the level of white blood cells. The results of early studies of the treatment of thalassaemia with hydroxyurea were not so impressive.

In 1985 Perrine and her colleagues noted that there was a delay in the switch from fetal to adult haemoglobin in infants of diabetic mothers and suggested that this effect may be mediated by the elevated level of butyrate in the diabetic patients. This suggested a quite different approach for attempting to elevate Hb F levels. In 1993 Perrine, Olivieri, and their colleagues reported the results of a short-term trial of arginine butyrate given to a patient who was homozygous for Hb Lepore (this form of thalassaemia was described in several earlier chapters). There was a dramatic response, with a major rise in both the haemoglobin and Hb F values. Because of the difficulties in long-term treatment with this agent, this patient was later switched to a combination of sodium phenylbutyrate and hydroxyurea. Her brother also had a similarly dramatic response to treatment with this combination, and both patients remained transfusion-independent for many years. Unfortunately, these agents were much less effective in the treatment of patients with β thalassaemia and its variants, and it has never been clear why one particular family responded so well; possibly they had different genetic constitutions with regard to fetal haemoglobin production, although, despite extensive searches, the mechanism was not discovered. Small studies on other groups of patients with Hb Lepore thalassaemia showed equally disappointing responses. Although these agents were used in combination with a variety of other drugs, including erythropoietin, and despite the undoubted success of the use of hydroxyurea for the treatment of sickle-cell disease, no further significant successes were achieved for the management of thalassaemia using these approaches.

Several other experimental approaches to treatment were tried. Once it became clear that the effects of the globin chains that are produced in excess in thalassaemia mediate much of their damage through oxidative mechanisms, a number of trials of anti-oxidants were established, none of which showed any effect on the survival of the red blood cells. Several studies were carried out to determine whether the hormone erythropoietin, which is produced by the kidney in response to anaemia and which stimulates red-cell production in the marrow, might, given in high doses, have a beneficial effect on the anaemia of severe thalassaemia. Although this never seemed likely, particularly since these patients have very high levels of this hormone anyway and their marrows are already overactive, several studies were carried out, but with no real benefit.

The Ultimate Goal: Gene Therapy

It is not surprising that, when the first information about the molecular basis of the thalassaemias started to emerge in the late 1970s, thoughts began to turn towards whether it would be possible either to replace defective genes or at least to modify them so that their activity was improved. The concept of gene therapy was widely discussed, and there was considerable optimism about its likely success. By the mid-1980s editorials were appearing in leading journals under headings such as 'Gene therapy around the corner'. In the event, it proved to be an extremely long corner.

Although by the early 1980s it was possible to isolate individual globin genes without too much difficulty, what had to follow to correct α or β thalassaemia posed serious questions: how could the gene be inserted into the appropriate blood cell precursors of a thalassaemic child, would the gene be regulated such that it produced appropriate amounts of its protein product in its new home, how long would it survive, what risks might arise from the manipulation of human DNA in this way, and many more. Essentially, there are two main approaches to gene therapy. First, there is somatic-cell gene therapy, in which an attempt is made to insert a gene into the appropriate tissue in which the defective gene is expressed. This gene, at least theoretically, remains in the tissue and for this reason cannot be passed on to the progeny of an individual treated in this way. The second approach, germ-cell gene therapy, is quite different. Here, the idea would be to identify that a newly fertilized egg carried a severe form of thalassaemia and then insert a normal globin gene to correct the condition. In this case the gene might be present throughout the individual's DNA and would be passed on to its progeny. All the early studies on gene therapy for thalassaemia focused on somatic-cell gene therapy.

The first attempt at gene therapy for severe thalassaemia was made by Martin Cline, an American haematologist. Having failed to obtain permission from an ethics committee in the USA to perform an experiment of this kind, Cline attempted to transfer a normal β globin gene into the marrow cells of a patient in Israel. The result was, not surprisingly, a total failure. Since at that time very little was known about the regulation of the globin genes, there were no efficient methods for gene transfer, and there were unexplored doubts about the possible dangers of the procedure, this attempt was viewed with deep concern by the scientific community. Fortunately, however, no harm was done to the patient, and at least some good came out of this episode. In particular, strict regulations for future work in this field were established in many countries. For example, concerned lest somebody else might attempt such a premature experiment in the United Kingdom, I approached the Medical Research Council, and a working group was established with all the research councils in Europe to draw up some preliminary guidelines for the application of research in this field. In 1989 this led to the UK government establishing a further committee under the chairmanship of Sir Cecil Clothier to review the whole field of gene therapy. A complete embargo on any form of germ-line gene therapy was advised, and rigorous conditions were to be met before somatic-cell gene transfer was attempted.

Following identification of the major regulatory regions in the β globin genes in the 1980s, the next problem was how it might be possible to transfer these genes into recipient red-cell progenitors. Although several physical approaches were investigated, none of them was able to transfect sufficient numbers of target cells. Thoughts therefore turned to nature's way of gene transfer. A series of studies were carried out to determine whether it might be feasible to use retroviruses—that is, viruses that are able to insert their genetic material into the genome—as potential vectors. The idea was to insert the required globin gene and to remove many of the viral genes so that none of their proteins is made in the cells that they infect. The virus's replication functions were provided by packaging cells—that is, cells that contain 'helper' viruses that produce all the viral proteins that are required, but that themselves have been disabled so that they cease to become infectious viruses. A great deal of work was directed at this ingenious form of genetic engineering, and some successes were recorded by the late 1980s, indicating that tissue-specific expression of a retrovirally inserted human β globin gene could be obtained in bone marrow or tissue culture. A number of other viruses were also explored as possible vectors, and a variety of other potential ways of introducing DNA into cells was also explored.

By now what was required of course was a good animal model of thalassaemia to test these various gene transfer vectors. This was essential, not only to determine whether the introduced genes would function normally and for a reasonable time, but also to attempt to assess the safety of the procedure. During the late 1980s and early 1990s, as the result of remarkable advances in genetic engineering, it became possible to modify the genomes of mice and to produce reasonable working models of human thalassaemia in these animals. The stage was now set to analyse the possibility of human gene transfer in a much more logical way. But innumerable technical problems were still encountered.

Interestingly, the first human genetic diseases to be treated successfully by these new approaches were immune-deficiency disorders. However, even this field encountered a severe setback early in the new millennium, when children treated in this way developed acute leukaemia of a rather unusual kind. The French scientists who did this work, much to their credit, dissected the sites of integration of the new genes and determined that they were close to a region where a particular gene, if abnormally activated, might lead to neoplastic transformation. Thus ways of preventing this problem in the future had to be worked out. Although there are many uncertainties at the time this book is being written, there are tentative reports of some success in trials of gene transfer therapy for thalassaemia.

During the same period other approaches to corrective gene therapy were explored. Some of them were directed at trying to correct a genetic defect—a splicing or nonsense mutation, for example—but, although these were successful in the test tube, the problem of scaling them up to a level to treat a patient was formidable. Another approach, again using nature's way of swapping genes, was to attempt what is called site-directed recombination. In principle this entails lining up DNA containing the defective gene with similar DNA with a normal gene and persuading them to change places. Again, this can be done reasonably efficiently in the test tube, but again the problem of scaling up the procedure sufficiently to cure a human disease is, at least up to the present time, insurmountable.

While at the time of writing this book it does appear as though there is a glimmer of hope that some form of gene therapy will become effective for the management of thalassaemia within the foreseeable future, it has taken over a quarter of a century of hard work by some extremely talented medical scientists to get this far. Indeed, apart from improved counselling and prenatal diagnosis, it has to be said that the application of molecular biology to this field has yet to make a major impact on the better management of the thalassaemias, a theme to which we shall return in the final chapter. However, as we have seen, the general level of treatment for children with thalassaemia slowly improved from 1970 to the new millennium. Unfortunately, this has not been the case for children in the developing countries where the disease is so common, a topic that we will consider briefly in the next chapter.

VIII

THE COMMONEST GENETIC DISEASES:

Was Haldane Right?

It has been apparent for many years that the inherited haemoglobin disorders must be the commonest diseases that are inherited in a Mendelian fashion—that is, the result of a single defective gene. In a survey in 2006 by the American charity March of Dimes it was estimated that approximately 7.2 million children are born with a congenital abnormality or a genetic disease each year. Of these numerous conditions, 25 per cent are comprised of only five diseases, of which the haemoglobin disorders are one of the two monogenic diseases, the other being glucose-6-phosphate dehydrogenase deficiency, another condition that appears to convey protection to malaria. The report also emphasizes that by far the highest frequency of all these conditions occurs in the countries with the lowest gross national incomes.

Particularly as most genetic diseases are, as would be expected, quite rare, the extraordinarily high frequency of the genes for the different forms of thalassaemia raises a number of intriguing questions: from where did these diseases arise, why did they achieve such a high frequency, how common are they, and what has been and is the lot of the hundreds of thousands of children with these conditions in the poor countries of the world?

The Origins of Thalassaemia

As evidenced by its name, for many years it was assumed that the thalassaemias had arisen somewhere in the Mediterranean region. An early thesis that gained considerable popularity was that the disease originated in an ancient race, Palaeoinsulara Mediterranea, which inhabited Sicily, Greece, and parts of Italy, from where it spread to other parts of the world. Several variations on this theme were also proposed. For example, it was suggested that thalassaemia was distributed in the sixth and seventh centuries BCE by Greeks who colonized Magna Grecia, including Sicily and southern Italy, and later spread eastwards within the empire of Alexander the Great. Speculations along these lines continued throughout the middle of the twentieth century, but in 1978 David Todd, in his inaugural lecture to the university of Hong Kong entitled 'Genes, Beans and Marco Polo', gave an absorbing and often extremely lurid account of the movements of the Mongoloid peoples, coming to the conclusion that it is likely that the thalassaemia genes had arisen in the east and followed these populations westwards.

This interpretation of events led to some rapid reconsideration of the origins of the thalassaemias. Views became divided between those who believed that thalassaemia had arisen in the Mediterranean and those who found Todd's new hypothesis more attractive. Yet others, sitting firmly on the fence, suggested that the disease had arisen somewhere in the Middle East and moved through migration both east and west! As it turned out, Todd's lecture and its subsequent publication were an unfortunate piece of timing. From the 1980s onwards, as the mutations were identified in patients with β thalassaemia from many different parts of the world, it became apparent that each country with a high frequency of the disease across the tropical belt had completely different sets of mutations. Clearly, the thalassaemias had arisen by independent mutations in different populations and then expanded by selection or other genetic mechanisms.

So it was that within a very short period in the 1980s all the previous hypotheses about the origin of the thalassaemia genes had been proved to be wrong. But many incorrect ideas that became well established in scientific or medical thinking over the centuries took a while to disappear from the literature, particularly if they had been voiced with enough conviction. When reviewing the *Cambridge World History of Human Disease* in 1994, I was dismayed to read the following description under the heading 'Thalassaemia': 'it is most frequent in areas where ancient Greek immigration was most intense.' Whether this is a reflection of the fact that the thalassaemia field was never very active in Cambridge or of the long gestation of the *World History* is not clear!

Why are the Thalassaemias so Common?

In an earlier chapter we discussed Haldane's hypothesis that the thalassaemias are particularly common in the Mediterranean region, because heterozygotes—that is, carriers—might have been protected from the malarial parasite by virtue of their small red cells. Curiously, Haldane's hypothesis was first tested, apparently without prior knowledge that it existed, by Anthony Allison in his pioneering work in Africa on the resistance to malaria of sickle-cell heterozygotes.

In 1949 Allison participated in the Oxford University Expedition to Mount Kenya. During this trip he collected blood samples from tribes all over Kenya for blood grouping and for studying other genetic markers, including the sickle-cell trait. He found low frequencies of the trait in populations living in areas where there was no malaria transmission and relatively high levels where there was a high rate of malaria infection. This suggested that the sickle-cell trait might be protective against malaria. Later he demonstrated that children with the sickle-cell trait have a lower level of parasites in their blood, and that adult sickle-cell heterozygotes inoculated with the malarial parasite *P. falciparum* did not show clinical evidence of infection as frequently as normal individuals. Although some of this work was controversial at the time, further evidence amassed slowly over subsequent years showed that he was correct. Indeed, detailed case-control studies in the period between the late 1990s and 2005 demonstrated that the sickle-cell trait provides a remarkable degree of protection against the severe complications of malaria.

Progress towards examining Haldane's hypothesis in relationship to the high gene frequencies for thalassaemia in many populations was much slower. As early as 1946, three years before Haldane proposed the malaria hypothesis, Vezzoso noticed that the distribution of thalassaemia in Italy is similar to that of malaria. This certainly tended to hold true in the extensive population surveys of the distribution of the gene in Italy by Silvestroni and Bianco, cited in an earlier chapter. The most important and widely quoted evidence for the association between β thalassaemia and malaria was first reported by

Carcassi and his colleagues in 1957 and later by Siniscalco and colleagues, based on population studies in Sardinia. Though the distinction between α and β thalassaemia was not fully established at the time of Carcassi's studies, it is almost certain that this work described the distribution of β thalassaemia in the island. In many ways Sardinia is an ideal population for this type of research, since it has many villages of great antiquity that have remained isolated for long periods. Furthermore, because of the geography of the island and the high altitude of some of these villages, there are malaria-free populations of similar ethnic background to the lower-lying village populations. Carcassi and Siniscalco demonstrated a clear correlation between the frequency of malaria and β thalassaemia in their studies in Sardinia. Like Allison's early findings in Africa, this work was subjected to heavy criticisms, in this case based mainly on uncertainties about the origins of the populations in the mountains and lower-lying villages, although in retrospect it is far from clear whether these concerns were justified.

Subsequent studies carried out during the 1960s failed to demonstrate correlations between β thalassaemia and malaria in Greece, Cyprus, Malta, Sudan, and New Guinea. In 1970 similar negative or extremely equivocal results were obtained in Thailand. One of the difficulties during this period was lack of knowledge about the extraordinary heterogeneity of the thalassaemias and of the tools required to distinguish between selection, gene drift, in-breeding, and many other mechanisms that can modify the distribution of particular genes within populations.

After a rather fallow period in the 1970s, the availability of the new information about the molecular basis of the thalassaemias, which started to become available at the beginning of the 1980s, led to a major revival of activity in this field. In 1986 my team in Oxford carried out extensive studies of the frequency of the mild form of α thalassaemia, α^+ thalassaemia, in the south-west Pacific region, a part of the world in which there was excellent information about the transmission rates of malaria. It was found that α thalassaemia has a clinal distribution—that is, a regular change in a biological trait, from north/west to south/east, with the highest frequency in northern New Guinea and the lowest in New Caledonia. These frequencies showed a strong correlation with malarial endemicity; a similar relationship was not observed for other genetic variants across this region. Of course, this gradual reduction in the frequency of α thalassaemia from north to south might have resulted from the gene being introduced from the Asian mainland by the original settlers and becoming diluted in their travels from north to south as they populated the different island groups. It was here that the real value of having information about the α thalassaemias at the molecular level became apparent. For, when the α globin genes of the island populations were examined, it was found that the form of α thalassaemia in Melanesia and New Guinea is quite different from that on the Asian mainland, and, what is more, it is set in a different genetic background, or haplotype. Clearly, this form of α thalassaemia had not been introduced from the Asian mainland but had arisen locally.

An unusual feature of the distribution of α thalassaemia in this region was that it was also found in Fiji in the west, Tahiti and beyond in the east, and in various Micronesian atolls, populations in which malaria has never been recorded. However, in these regions almost 100 percent of α thalassaemia could be accounted for by the same mutation that was identified in Vanuatu, indicating that the occurrence of α thalassaemia in these non-malarious areas had been the result of population migration and founder effects.

These convincing population data were further augmented by prospective case-control studies in northern Papua New Guinea, where it was found that, compared with that of normal children, the risk of being admitted to hospital with severe malaria, as defined by the World Health Organization guidelines, is 0.4 for α^+ thalassaemia homozygotes and 0.66 for α^+ thalassaemia heterozygotes. So these conditions offer a 40–60 per cent protection against the likelihood of being admitted to hospital with the severe forms of malaria.

These studies, carried out between the mid-1980s and the late 1990s and confirmed later in other populations, provided extremely strong evidence that the central tenet of Haldane's malaria thesis was correct—that is, that the extremely high frequencies of the mild forms of α thalassaemia that occur throughout sub-Saharan Africa, in the Middle East, and throughout Asia are the result of the relative resistance of heterozygotes, and in this case homozygotes, to the most severe form of malaria, that is *P. falciparum* malaria. While case-control studies of this type were not carried out for β thalassaemia or Hb E over this period, further circumstantial evidence was obtained that they too are malaria-resistant polymorphisms.

When we discussed Haldane's malaria hypothesis in an earlier chapter, he was quoted as suggesting that the reasons for the potential protection against malaria by carriers for thalassaemia might result from the small size of their red blood cells, which, he suggested, might provide an unsatisfactory environment for the malarial parasite. Was he also correct in this aspect of his hypothesis?

Over many years, research in the malaria field was hindered because of lack of an experimental system in which to study the parasites that produce the disease. However, in 1977 James Jensen and William Trager developed a method for the *in vitro* culture of malarial parasites. Unlike the reliance on high technology that has led to most of the advances described in this book, this technique was refreshingly simple; since the parasites seemed to thrive in an atmosphere of reduced oxygen, the culture plates on which they were grown were placed in a jar together with a lighted candle! This made it possible to grow the most important human pathogen, *Plasmodium falciparum*, and to compare its rates of invasion and growth in both normal human red cells and in red cells obtained from carriers for different forms of thalassaemia or haemoglobin variants. Work in a number of laboratories over the next few years showed quite unequivocally that the rates of invasion and development in the cells of carriers for α or β thalassaemia are no different from those in normal red cells. Clearly, Haldane's notion that the small red cells characteristic of these carrier states would be an unattractive home for malarial parasites would not hold up. Later, and at least in the case of the α thalassaemias, it was found that particular properties of the membranes of the red cells may account, at least to some extent, for their protective effects against malaria. But the final story of how these protective effects are mediated remains to be told.

Haldane was not entirely wrong, however. As we have seen in earlier chapters, Hb E, because it is synthesized at a reduced rate, is characterized by a mild β thalassaemic phenotype.

In vitro studies of carriers or homozygotes for this variant have suggested that their red cells are somewhat resistant to both invasion and growth of malarial parasites. Interestingly, the availability of *in vitro* culture techniques was much more effective in pointing the way to the mechanism of protection against malaria by the sickle-cell trait. It was found that the rate of sickling of parasitized red cells in those with this trait is significantly greater than that of non-parasitized cells. Furthermore, it was also demonstrated independently by workers in Oxford and New York that parasitized cells from those with the Hb S trait maintained at low oxygen tension do not support the growth of malarial parasites as effectively as normal cells under the same conditions. These findings offered at least one reasonably clear-cut mechanism for the protective effect of the sickle-cell trait against malaria.

It is beyond our scope to describe in detail how the developments in cell and molecular biology combined to provide a much clearer picture of the reason for the remarkably high levels of the haemoglobin variants, a field that evolved dramatically in the early part of the new millennium. The picture that emerged is that disorders like the thalassaemias occur at very high frequencies in malarious areas, or those that have been exposed to malaria in the past, and that in each region different molecular forms of thalassaemia are found. These observations, and the detailed analysis of the genetic background or haplotypes on which these mutations occur, suggested that they have arisen relatively recently in evolutionary terms. Although the ancestral forms of the varieties of malaria that affect humans may have arisen from ancestors in Africa many million years ago, it is likely that *Plasmodium falciparum* arose from its closest ancestor approximately ten million years ago.

After ten million years of development in Africa, modern humans emigrated out of the continent some 40,000–100,000 years ago, probably carrying some of the polymorphisms that evolved from selection against malaria. It is likely that death due to malaria increased dramatically between 5,000 and 10,000 years ago, with the development of agriculture and settlements that would have greatly facilitated the transmission of the disease by mosquitoes. These observations are all compatible with the fairly recent appearance of malaria-resistant polymorphisms, and they go some way to explaining the absence of thalassaemia in the indigenous populations of the New World. Presumably the selective factors leading to their high frequency in Asian populations had not been present before their early migrations into the New World. It is still not certain when malaria reached this region, although it has been suggested that it may have been as recently as the early Spanish Conquests. Hence there may not have been time for selection to generate high frequencies of malaria-resistant polymorphisms like the thalassaemias.

How Common are the Thalassaemias?

As discussed briefly in previous chapters, isolated reports of the occurrence of thalassaemia in countries right across the tropical zones started to appear in the literature after the Second World War. In 1959 Chernoff wrote a review of what was known at the time of the world distribution of thalassaemia, which included the first map of the disease. When this review was written, with the exception of some of the Mediterranean populations, very little was known about the carrier frequency, and the map simply described regions with foci of thalassaemia genes.

Examining hundreds of samples to detect the carrier state for thalassaemia is no mean task and required the development of simplified and inexpensive methods. In the 1940s Silvestroni and his colleagues in Italy developed a very simple test based on the principle that the red cells of thalassaemia carriers are more resistant to lysis in dilute solutions of saline than normal cells. Thus, if a saline solution is made at a particular concentration such that normal cells will lyse and thalassaemic cells will not, it is possible to test for the thalassaemia trait by simply placing a drop of blood in the solution and determining whether a clear red haemoglobin solution forms, indicating that the cells are lysed, or whether the solution remains cloudy because of a suspension of unlysed cells. This approach was used widely in many of the early population studies for the frequency of the thalassaemia trait. Later, more sophisticated analysis of the size of the red cells combined with haemoglobin electrophoresis took the place of the one-tube test.

Some of the earliest and largest surveys were carried out in Italy by Silvestroni, Bianco, and their colleagues between 1944 and 1981. These early studies showed that the distribution of the disease was extremely variable in different parts of Italy, with the highest frequencies in the Po delta in the north and in the southern regions of Campania, Calabria, Puglia, and Sardinia. In the Po delta the incidence of carriers ranged from 7 per cent to 19 per cent. Other surveys carried out at about this time in Italy observed the same unevenness of distribution of the carrier rates. For example, in 1961 Brancati studied 33,000 people in 121 areas of Calabria. There was a remarkable variation in the frequency of the disease, from less than 4 per cent to over 9 per cent, reaching just under

20 per cent in some regions. The carrier rates seemed to be higher in low-lying areas, particularly on the coast of the Ionian Sea. During the 1960s high but extremely variable frequencies of the thalassaemia trait were also found in Greece, the Mediterranean islands, and in some areas of East and Central Europe.

Some of the earliest surveys in India were carried out by Chatterjea, who, in 1959, found a carrier frequency of 3.7 per cent for β thalassaemia trait in Bengali populations together with a high preponderance of patients who had inherited both Hb E and β thalassaemia. Over the next ten years innumerable small studies were carried out in different areas of India, showing that the thalassaemia gene is widespread throughout the Indian subcontinent, although the relative frequency varied widely even within short geographical distances. Between the late 1950s and 1970 similar data started to appear from many countries in South and South East Asia, and there were preliminary reports suggesting that the disease occurs at a high frequency in many parts of Southern China. At the same time it was becoming clear that Hb E, which is in effect a mild form of β thalassaemia, is extremely common in Northern India, Burma, Thailand, Malaysia, and Indonesia, reaching remarkable frequencies of up to 70 per cent in the border region between northern Thailand and Cambodia, an area that later became known as 'the Hb E triangle'. It also became apparent that thalassaemia major was the most serious form of the disease in the Mediterranean region, the Middle East, and many parts of the Indian subcontinent, while the co-inheritance of Hb E and β thalassaemia, Hb E β thalassaemia, was the most common form in eastern parts of India, Bangladesh, Burma, Thailand, and Indonesia. Later studies suggested that there

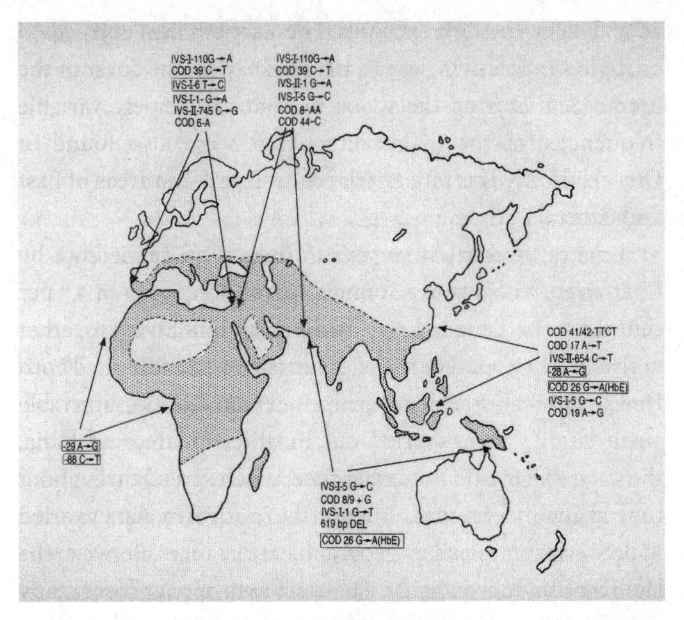

17. The world distribution of β thalassaemia and the common mutations in each region. The mutations that are boxed cause a particularly mild reduction in β globin production. Hb E is produced at a reduced rate and results in the clinical picture of mild β thalassaemia when inherited in the homozygous state; when inherited together with β thalassaemia, it causes a condition called Hb E β thalassaemia, which is the commonest form of serious thalassaemia in the region east of India. COD stands for codon—that is, three nucleotide bases. IVS stands for intervening sequence. DEL represents a gene that is deleted or partially deleted—for example, 619bpDEL means that 619 base pairs have been removed from the gene.

might be as many as 100,000 patients with this condition in Thailand and twice this number in Bangladesh. The world distribution of β thalassaemia, Hb E, and the common mutations that cause β thalassaemia are shown in Figure 17.

Until the molecular era, it was not possible to obtain anything approaching an accurate determination of the frequency of α thalassaemia except by surveys of umbilical cord blood samples for measurement of the level of Hb Bart's. From the early 1980s onwards surveys were initiated to examine the frequency of the two common forms of α thalassaemia at the DNA level. Remarkably, it was found that the mild form of α thalassaemia, α^+ thalassaemia, which is due to the loss of one of the two linked α globin genes, occurs at an extremely high level, in the 15–30 per cent range, across sub-Saharan Africa, the Middle East, the Indian subcontinent and throughout East and South East Asia. In some regions of north India and Papua New Guinea the carrier frequencies for this condition approach 70 per cent or 80 per cent of the population. On the other hand, the severe form of α thalassaemia, α^0 thalassaemia, was found only at high frequencies in parts of South East Asia and the Mediterranean islands. Clearly, therefore, the homozygous state for this condition, the Hb Bart's hydrops syndrome, was only likely to be encountered in these regions (Figure 18).

In 1967 and 1985 the American geneticist Frank Livingstone produced two remarkable works, summarizing what was known about the frequencies of the haemoglobin disorders in every country. A review of the literature of the field from the mid-1980s onwards suggests that, while the availability of DNA technology had made it easier to assess the frequency of the α thalassaemias, it also had the effect of making extensive population surveys for the gene frequencies of the thalassaemias a less attractive pastime. It was, it appears, much more exciting to define the different thalassaemia mutations! Hence, and particularly in view of the extraordinary diversity of the gene frequencies of the thalassaemias even within small geographical distances, a true estimate of their frequency and potential health burden remained unclear.

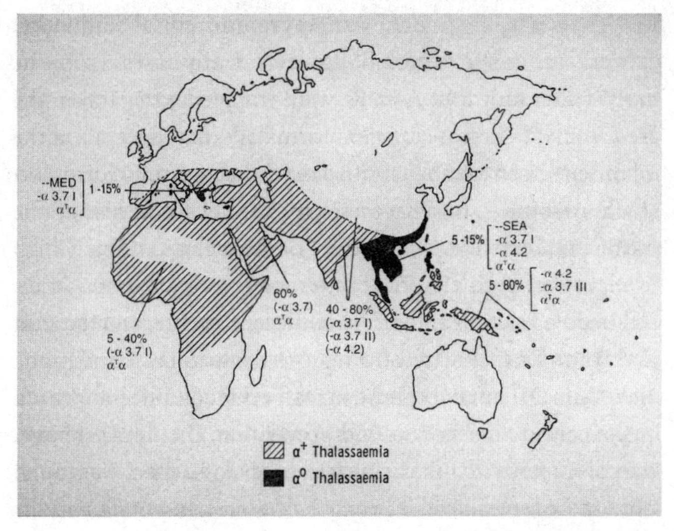

18. The world distribution of the α thalassaemias. The figures give an approximate indication of the frequency of the different forms of α thalassaemia in the population. α^0 thalassaemia is restricted to South East Asia and some of the Mediterranean islands. The various subheadings, −α3.7 I, II, or III, reflect different molecular forms of α^+ thalassaemia, $—^{MED}$ and $—^{SEA}$ describe different size deletions, which remove both linked α globin genes in patients from the Mediterranean or South East Asia region, respectively.

In a series of workshops and publications between 1983 and 1994, the World Health Organization (WHO) attempted to estimate the frequency of the α and β thalassaemias together with the common haemoglobin variants throughout the world. The problem with these publications is that the frequency data are not all referenced and in fact some of them reflect personal communications from individual centres in different countries. Given the extraordinary heterogeneity of distribution of the disease, it is clear that these data have to be accepted with extreme caution. In the last of these reports, it was estimated that there are approximately 270,000,000 carriers for the haemoglobin disorders, many of whom live in the WHO South East Asia Region, where the thalassaemias predominate. It was further estimated that over 300,000 infants are born each year with a major haemoglobin disorder, of whom some 70,000 have a severe form of thalassaemia and the remainder sickle-cell anaemia or its variants.

Very few countries have attempted to carry out more detailed population studies since the early 1980s, and the true global burden that results from the thalassaemias still remains uncertain. Of course, because of major population movements in the period after the Second World War, the thalassaemias were also seen with increasing frequency in all the richer countries of the world. And it became clear that, although natural selection had been the major basis for the extraordinary high frequency of these genetic diseases, further factors, particularly the common practice of consanguineous marriage, were also playing an important role.

Thalassaemia in Developing Countries

The slow improvements in the management of the different forms of thalassaemia that occurred during the second half of the twentieth century in many of the richer countries had little impact on patients in many parts of the developing world. This is not surprising, since many of these countries continued to suffer from the ravages of malnutrition, major infectious killers, inadequate sanitation, and poorly developed health-care systems. In this environment many babies born with thalassaemia died of other causes before the disease was recognized. However, as at least some of these countries went through the epidemiological transition following improvements in sanitation and public health care, these babies began to survive long enough to present for diagnosis, a phenomenon that we touched on in a previous chapter when we discussed the extraordinary increase in the number of patients with thalassaemia in Cyprus after the Second World War.

It was undoubtedly these remarkable demographic changes in these common genetic diseases that, from the 1950s onwards stimulated the development of partnerships between centres in the richer countries with expertise in thalassaemia and those in the poorer countries where knowledge of the disease was extremely limited; in global health jargon these interactions became known as North/South partnerships. We saw in an earlier chapter how the partnership between the haematology group in St Louis with Thailand led to the development of expertise in this field very early during its development in that country. Similar partnerships between centres in the UK and elsewhere developed during the 1970s, some of which were relatively short-term training programmes while others lasted for many years. This approach led to the development of genuine centres of excellence in the thalassaemia field in many countries, including Cyprus, Sardinia, parts of India, and others. Unfortunately, however, many of the poorer countries, particularly those of Asia, at least in part because of the preoccupation of their governments with disease caused by malnutrition and infection combined with natural disasters and war, were not able to develop any programmes for the control and management of thalassaemia, a disease that was completely neglected. Early in the new millennium thoughts turned to developing South/South partnerships between countries where there was expertise in the control of thalassaemia and those where none existed. But in no small part because of the lack of interest in non-communicable disease as a whole, and genetic disease in particular on the part of global health agencies, it proved difficult to fund such partnerships, although slow progress was made.

Clearly, therefore, the thalassaemias and other common disorders of haemoglobin are by far the commonest single-gene disorders, and, as judged by the World Health Organization, about 7 per cent of the current population of the world carry a gene for one or other of these conditions. This appears to have been the result of a remarkable example of natural selection whereby, as the different mutations that cause these diseases arose, they rapidly expanded, because carriers were more resistant to severe forms of malaria. Although consanguinity may also have played a role in increasing the gene frequency of these conditions, their present numbers today seem largely to reflect the effects of Darwinian evolution at work. The remarkable variability in the distribution of these disorders even within relatively short geographical distances can often be related to the presence of malaria, either now or in the past, together with population movements and founder effects. Clearly, research on this disease carried out over the second half of the twentieth century not only paved the way for the beginnings of an understanding of genetic disease at the molecular and cellular level, but also emphasized the important role of evolutionary pressures in generating the current pattern of human disease. Unfortunately, this mechanism for the generation of extremely high frequencies of the different forms of thalassaemia meant that the diseases occurred at their highest frequency in parts of the world where poverty and ill health were rife and where it was extremely difficult to develop the medical services required for the control and management of the thalassaemias. Although conditions improved in some countries, at least in part as a result of the formation of North/South partnerships, for many this was not the case, and, because of the excusable neglect on the part of their governments and the less excusable neglect on the part of international health agencies, the lot of a baby with thalassaemia was little better than it had been in the first half of the twentieth century.

IX

EPILOGUE: THALASSAEMIA AND MOLECULAR MEDICINE

As we have seen as we surveyed eighty years of the evolution of knowledge about the thalassaemias, developments in this field from the 1950s onwards were set against the background of an extraordinary period of progress in the biological sciences. It was soon clear to those who entered the field at that time that the remarkable advances in molecular and cell biology were likely to have important implications for a better understanding of disease mechanisms and even for their more effective management in the future. There was a slow shift in emphasis in medical research from studies in the community and at the bedside to the characterization of disease at the cellular and molecular level. Somewhere along the way the term 'molecular medicine' came into general, if premature, usage. Hence, in rounding up our story, it is interesting to examine the role of research in thalassaemia in the evolution of this new field and, perhaps more importantly with our patients in mind, to ask what genuine advances stemmed from this new concept for the benefit of those with different forms of thalassaemia in particular, and other diseases in general.

Thalassaemia and the Evolution of Molecular Medicine

As we saw in Chapter 3, the early development of the thalassaemia and abnormal haemoglobin field evolved in the 1950s and 1960s, at least in part through the chance coming-together of scientists from different disciplines. Because haemoglobin, after insulin, was one of the first proteins to be analysed in detail at the structural level, and with the discovery of abnormal haemoglobins in 1949, the field was activated by several outstanding protein chemists and scientists from related fields, together with a small number of human geneticists and haematologists who had learnt enough at least to be able to converse with their protein-chemist colleagues. Many of these scientists were well aware of what was going on in the evolving field of molecular biology, and, although most of them moved later into other fields, they remained in the haemoglobin field long enough for the scene to be set for the analysis of these diseases at the molecular level.

The other major factor in the early evolution of the thalassaemia field at the molecular level was, of course, the relative ease of studying blood diseases compared with other genetic disorders. Haemoglobin in red blood cells is, in effect, served up by nature ready for study; when you cut your finger and wash it under the tap, the red solution that goes down the drain is almost 100 per cent pure haemoglobin. For the protein chemists of the 1950s, who might have to spend months purifying trace amounts of a liver enzyme, to have the particular protein that they were studying already purified for them in this way was a great incentive to work in this field. Similarly, in the DNA era the young red cell was a ready source of almost pure α and β globin messenger RNA for the development of DNA probes. Given these technical advantages, together with increasing awareness of a diverse set of inherited diseases in which abnormalities of the blood seemed to be a major factor, it is not surprising that the haemoglobin disorders were the first conditions to be explored at the molecular level.

As we saw in earlier chapters, the haemoglobin field was developing so rapidly that as early as the mid-1960s and early 1970s at least two clear-cut molecular mechanisms underlying the thalassaemias had been anticipated at the protein level, and, once the globin genes had been isolated at the beginning of the 1980s, within only a few years a remarkable spectrum of the different types of mutation that must underlie genetic disease had already been determined. Similarly, the discovery of linkage between the haemoglobin loci and restriction fragment length polymorphisms was, in effect, proof of principle that it should be possible to search for genes for other diseases using the linkage approach. Indeed, it was applied by geneticists in other fields within a few years after its discovery.

One of the other major developments of importance in the early days of thalassaemia research was the realization that a so-called simple monogenic disease like thalassaemia is by no means 'simple', in the sense that the associated clinical findings are extremely variable and that at least some of this clinical diversity is due to the effect of other genes, or modifiers.

There is another and often neglected aspect of the contributions of the haemoglobin field to the early development of molecular medicine and to a better understanding of the function of genes in human beings. Throughout the evolution of this field of research it became apparent that the study of unusual haemoglobin variants or forms of thalassaemia often shed considerable light on the processes of normal physiological function. For example, the discovery of haemoglobin variants with abnormal oxygen-carrying properties or decreased stability provided critical information about the mechanisms of oxygen transport and haemoglobin stability in health. And a detailed analysis of the different mutations that underlie rare forms of thalassaemia provided invaluable information about the ways in which the synthesis of haemoglobin is regulated. In fact, nearly all the major regulatory regions in the globin gene clusters were discovered from the study of patients with unusual forms of thalassaemia in which their function was defective.

Undoubtedly, therefore, the exploration of the thalassaemias and abnormal haemoglobin disorders at the protein and DNA level opened the door to the study of monogenic disease at the molecular and cellular level, shed early light on the regulation of genes, and, as described in the previous chapter, opened up new avenues for the exploration of human evolution. But its overall influence on the later development of molecular medicine is more difficult to assess. For example, mouse models of cystic fibrosis were developed in the early 1990s. In 1995 it was discovered that there is a gene in mice that modifies the clinical picture of cystic fibrosis, a finding that was followed by banner headlines in the leading scientific journals, announcing that simple monogenic disease is not so simple after all! The existence of modifiers of this type had been well established in the haemoglobin field for close on thirty years, yet these findings were never mentioned or discussed in these publications, or in subsequent reviews of genetic modifiers of human monogenic disease over the next fifteen years. Nor did the molecular pathology of the haemoglobin disorders figure very highly in the literature of human genetics over this period. And, when thoughts turned to trying to dissect the complex interplay between different genes and the environment that underlie common diseases such as heart disease and diabetes, the lessons learned from the multilayer complexity of the genetic and environmental modifiers of less complex monogenic diseases like the haemoglobin disorders were never referred to. Why is this?

Although it is probably too early to assess the evolution of molecular medicine at the turn of the twentieth century, it is possible to hazard a guess as to why the haemoglobin field may have become sidetracked from the main stream of the development of human genetics over this period. Medical genetics evolved as a distinct specialty only in the period after the Second World War, and mainly in the USA and Europe. Since it developed in the richer countries, it is natural that it focused on chromosomal abnormalities and other genetic diseases that were relatively common in their populations. Although in many of these countries there were large numbers of immigrant families with children with thalassaemia or abnormal haemoglobin disorders, they tended to be looked after in haematology departments and were not seen commonly in the evolving medical-genetics units. Hence the haemoglobin field became separated from the main stream of medical genetics early in its development. Furthermore, the bulk of research literature of the haemoglobin field appeared in haematology journals; there were relatively few papers published on this topic in the increasing numbers of journals that covered medical genetics during this period.

This interpretation of the sidetracking of the haemoglobin field is strongly supported by the published version of Paul

Polani's Harveian Oration to the Royal College of Physicians, London, given in 1988 and published in 1990. In this lecture, entitled *The Impact of Genetics on Medicine*, in a table that lists the estimated incidence at birth of some relatively common and better-known recessively inherited genetic diseases, the haemoglobin disorders, which by then were well known to be by far the commonest diseases of this type, are not mentioned. Nor, with the exception of one reference to prenatal diagnosis, are they mentioned anywhere else throughout this extensive review of the field. It is possible, of course, that this review of the current state of medical genetics may have been partly influenced by Polani's particular interest and, in particular, by the kind of patients that were being referred to the rapidly evolving departments of medical genetics in the UK at the time. Interestingly, its omission of any mention of the haemoglobin field is in contrast to the account of human genetics published in one of the first extensive textbooks of the field, *Human Genetics*, edited by Vogel and Motulsky. Here, under the heading 'Hemoglobin as a Model System', the authors write: 'Hemoglobin is the best-analyzed genetic system in humans. Therefore experiences and concepts developed in the course of its analysis may help towards a better understanding of the phenomena in other fields of human genetics.'

Perhaps it was inevitable, therefore, that, given the rapid increase in medical specialization and the increasing complexities of genetic disease that were being unearthed during the middle of the twentieth century, the haemoglobin field moved away from the mainstream of medical genetics. It is too early to assess whether this had a deleterious effect on the pace of the evolution of knowledge about genetic diseases, but this separation certainly makes it more difficult to assess the overall role of early work in thalassaemia on the later development of molecular medicine. Interestingly, very few of the pioneers in the haemoglobin field moved later into the mainstream of medical genetics. And, although some of my British colleagues in medical genetics tell me that my small monograph, first published in 1982, *The New Genetics and Clinical Practice,* which used the haemoglobin disorders as its main model, encouraged them and others to move towards the field of human molecular genetics, I doubt if the overall effect was significant! The origins of scientific developments are often difficult to trace, however. Alec Jeffreys, the scientist who developed DNA fingerprinting, which is now used all over the world for catching criminals, paternity testing, and in many other fields of human genetics, told me that he first developed his work based on a highly variable DNA sequence that we had found near the human α globin genes, which he first saw in this book, after which he noted a publication of a related sequence near the insulin gene and then discovered a third sequence of this kind near the gene for myoglobin, the protein involved in oxygen transport in muscle. What followed is now well known, particularly to the criminal fraternity of the world. But it will be for historians of the future to try to sort out the genuine origins of what became rather hopefully known as 'molecular medicine'; the haemoglobin disorders undoubtedly helped to set the scene, but their later influence remains unclear.

Has Molecular Medicine Improved the Lot of Patients?

As we have seen, it is too early to assess fully the development of molecular medicine, let alone its effect of improvements in health care. From its earliest days, in the 1970s, there was a feeling of considerable optimism that an understanding of disease at the molecular level would rapidly result in improvements in its control and management. These expectations reached a climax in 2001 following the partial completion of the human genome project. At this time there was wide speculation among at least some scientists and in most of the media that this remarkable achievement, after which the human genome was described as the *Book of Life*, would change the pattern of medical practice as we know it within twenty years. Unfortunately, as we approach the halfway stage of this prediction, it is becoming increasingly clear that such optimism was premature.

Apart from the undoubted improvements in the diagnosis and prenatal detection of the thalassaemias, almost thirty years after the first discoveries of their molecular basis this new knowledge has had limited effect on their clinical management. And the same is true of most genetic diseases. For example, cystic fibrosis, a relatively common genetic disease in north European populations, is another condition for which the defective gene was discovered early in the development of molecular medicine. Knowledge of the many different mutations that can cause this condition, while it has led to a better understanding of some of its varied clinical manifestations, has had little effect on improving the lot of affected patients. The same applies to the discovery of genes that make it more likely that people will develop common diseases such as heart disease, stroke, and diabetes, although, again, these discoveries are helping to clarify the pathological mechanisms that underlie these conditions. Granted, the discovery of the molecular basis for rare forms of monogenic hypercholesteraemia by Brown and Goldstein in 1974 shed important light on normal cholesterol metabolism, and contributed to the development of cholesterol-lowering drugs, but such examples are rare. Undoubtedly the most remarkable applications of molecular and cell biology in medical practice have come from the cancer field, where they have completely revolutionized our thinking about the underlying causes of cancer and, indeed, have led to the discovery of a number of agents that are effective in its management.

It should be emphasized, however, that, while the pace of the generation of new therapeutic options based on molecular medicine was slow, the same cannot be said of the value of the technologies of this new field for the development of new diagnostic agents. As well as their value in the molecular and prenatal diagnosis of genetic diseases like thalassaemia, they have had a major influence on many other fields of medicine. For example, for infectious diseases, particularly those caused by organisms that are difficult to culture, they have proved invaluable for early diagnosis and more accurate classification. In particular, they have led to the rapid identification of the organisms involved in new epidemics and of potential targets for vaccine development.

Although the basis for the relative paucity of early therapeutic successes arising from molecular medicine is complex, and must be left for historians of the future to dissect more fully, at least one reason is clear. It is simply the time scale involved. Throughout history there has always been a long delay between fundamental observations in the laboratory and their application for the better management of patients. The story of the slow evolution of knowledge about infectious disease is a prime example. While the pace of development of the biological sciences advanced much more rapidly than at any other time in the latter half of the twentieth century, it was inevitable that the completely new information that was arising from studies at the molecular and cellular level would not find application in the ward overnight. So far, the fruits of the human genome have unearthed layer upon layer of complexity of normal biological systems, and have only begun to scratch the surface of the infinitely more complex interactions between genome and environment that underlie the problems of sick people.

Historians of the future will also want to understand why the time scale for the medical applications for the human genome project was so wildly underestimated and, incidentally, the effect that this may have had on the direction of medical science in subsequent years. There is no doubt that the successful completion of the human genome project was one of the most remarkable achievements in the biological sciences. But the degree of hyperbole about its potential for altering the whole pattern of medical practice was out of all proportion. After the announcement of the partial completion of the project, a few of us were invited to Downing Street to meet the then Prime Minister, Tony Blair. As he walked into the room to greet us, an enormous television screen lit up at the other end of the room, and there was President Bill Clinton together with Francis Collins and Craig Venter, who had been responsible for the public and private programmes for the American input into the genome project, respectively. The conversation between the two leaders was restricted to a discussion about how the health of Mrs Blair's new baby would benefit from the fruits of the genome project, such that it might live for over 100 years! Expectations for the fruits of the project were equally expansive in the media at the time, but undoubtedly some of this mood of optimism was fanned by the scientists themselves.

It should be remembered that, particularly in the early days of the evolution of molecular medicine, many workers in the field were scientists, who had limited knowledge of the complexities of sick people. It is not surprising that they believed that, once the cause of a disease was established, a cure would follow rapidly. And, as in most fields of science that are undergoing major expansion, research in molecular medicine became increasingly competitive. Hence there was a tendency on the part of the scientists involved to overstate the potential practical outcomes of their work to increase the likelihood of obtaining funding, a tendency that was encouraged by science journalists who were interested only in 'major breakthroughs'! However, the gradual realization on the part of governments and funding agencies that the fruits of this remarkable field for the betterment of patient care might be slower in coming than had been suggested was to lead to some major rethinking about priorities and organization of support for medical research.

A little over six years after the announcement of the partial completion of the human genome project, a new term, 'translational medical research', appeared on the scene, intimating that the organization and funding of medical research should be fundamentally changed in order to enhance its effect on patient care. While some of the thinking along these lines may have reflected concern about the balance of support for basic as compared with clinical research, it implied that medical scientists are more interested in science for its own sake than in the process whereby its fruits can be applied in clinical practice. But there is little evidence from a survey of the development of medical knowledge that this is the case. The isolation and therapeutic application of penicillin by a group in Oxford at the beginning of the Second World War followed years of curiosity-driven science about the properties of naturally occurring antibacterial agents; when an agent of this type was discovered during a survey of the literature, an enormous effort was made straightaway into exploring its therapeutic possibilities. Indeed, in this brief account of the history of the thalassaemia field, it is quite remarkable how discoveries such as ways of measuring globin chain synthesis in the test tube or identifying mutations in the globin genes were, within no more than a few years, finding applications in the clinic for prenatal diagnosis. And at every stage in the development of knowledge about the thalassaemias, efforts to apply new findings to the better management of the disease were a constant theme. The same applies to most research in the early days of the molecular era.

There is no evidence that the clinical applications of medical research can be hastened by the way that it is organized or even funded. Shortly after the American moon landing, the then President of the United States, possibly with an imminent election in mind, insisted that a large amount of money should be put into the sickle-cell field to discover how to cure sickle-cell anaemia, a disease that affects thousands of African-Americans. The result was some useful progress towards a better understanding of the mechanisms of disease, but the overall outcome was of no value whatever for its management. Research into most complex diseases takes many years of fundamental work in the laboratory before that moment when it is ready for exploitation in the clinic. It is only then that its application can be enhanced by increased funding and appropriate translational organization.

Since the major killers in the twenty-first century are likely to be diseases that reflect the complex interplay of many genes with the environment, set in the background of the ill-understood biology of ageing, their control will surely require a broad-based research approach ranging from fundamental biological science, through clinical investigation, to the social and behavioural sciences and epidemiology. These principles have been reflected clearly in this short account of the history of thalassaemia, a disease that is dubbed a 'simple monogenic disorder', since it is caused by a single defective gene. Yet the diverse clinical manifestations result from layer upon layer of interactions with other genes combined with variation in adaptation to the disease, the effects of the environment, and the evolutionary complexities that have made it so common. Added to all this, there are the equally complex social, religious, and ethnic factors that combine to determine the success of its control in different parts of the world.

In the case of the thalassaemias, like many other diseases, there is every reason to believe that further advances in cellular and molecular biology may, in the long term, play a major role in their better treatment and even their cure. There are already hints that this may be possible through one or other form of gene therapy, for example, but, if the history of medical and scientific advances over the last fifty years has told us anything, it is that it is unwise to try to predict when this may happen.

GLOSSARY

ALLELES alternate mutations at the same genetic locus

ALLOSTERIC an effect in which a molecule binds to one site on a protein, causing a change in its shape such that the activity at another site on the protein is altered

AMINO ACID small molecules that form the building blocks of proteins

ANAEMIA a reduction in the haemoglobin level in the blood below the population norm

BASE a substance with an alkaline reaction; the bases that form nucleic acids are called purines (adenine and guanine) and pyrimidines (cytosine, thymine, and uracil)

BASE SUBSTITUTION substitution of one base for another in the DNA

BONE MARROW the site of production of the blood cells in the central cavity in bones

CODON a triplet of nucleotides in messenger RNA, coding for a particular amino acid

CHELATING AGENT a chemical that is able to bind metals

CHROMOSOME thread-like bodies found in the nucleus of cells and containing the genes

COMPLEMENTARY DNA (cDNA) DNA that is copied from messenger RNA templates and used for probing the genome

DEOXYRIBONUCLEIC ACID (DNA) a polymer consisting of two strands of nucleotide bases in a double-helical configuration in which the sugar residue is deoxyribose; there are four bases, adenine (A), guanine (G), cytosine (C), and thymine (T); the two strands of nucleotide bases are held together by weak chemical bonds, A paired with T and C with G; because of these rules of base-pairing, when the two strands of DNA divide, each acts as a template for a new strand that has exactly the same base-pair composition

DOMINANT INHERITANCE allele manifesting its effect in a heterozygote; in medical genetics this is reflected in diseases that pass directly from one generation to another

DYSERYTHROPOIESIS abnormal production of red-cell precursors

ERYTHROPOIESIS the production of red blood cells

ERYTHROPOIETIN a hormone produced by the kidney in adult life and in the liver during fetal development that stimulates the production of red blood cells

EXON the coding sequence of a gene

FRAMESHIFT MUTATION the insertion or deletion of one or more bases in DNA that leads to scrambling of the genetic code

GENE the region of a chromosome with detectable function that reflects the particular order of nucleotide bases in a DNA strand

GENE DELETION the loss of all or part of a gene

GENETIC CODE the order of bases in DNA that underlies the function of genes

HAEM a ring structure that is bound to a globin chain and that carries the iron necessary for oxygen transport

HAEMOGLOBIN the oxygen transporter of the red blood cells

HAEMOGLOBIN A the variety of haemoglobin in adult red blood cells

HAEMOGLOBIN A_2 the minor haemoglobin of adult red blood cells

HAEMOGLOBIN F the haemoglobin of fetal red blood cells

HAEMOLYTIC ANAEMIA anaemia that is due to a shortened survival of red blood cells

HETEROZGYOTE different alleles of a particular gene, or locus, on homologous chromosomes

HOMOLOGOUS CHROMOSOMES chromosomes that carry genes governing the same characteristics and that pair together when cells divide

HOMOZYGOUS carrying the same allele at a particular locus on both of a pair of homologous chromosomes

HYPOCHROMIC ANAEMIA anaemia characterized by poorly staining red cells reflecting deficiency in haemoglobin

INITIATION CODON a region of messenger RNA that encodes a position for the beginning of peptide-chain synthesis

INTRON a sequence of nucleotide bases inserted into the coding region of a gene that has no informational function

LINKAGE the presence of two or more gene loci on a particular chromosome resulting in a tendency for alleles at these loci to be inherited together

LYSATE the contents of a cell that is released when its membrane is disrupted

MESSENGER RNA (mRNA) ribonucleic acid that is synthesized using a single strand of DNA as a template and that is then used to direct protein synthesis on ribosomes

MICROCYTIC ANAEMIA anaemia associated with small red blood cells usually indicating a reduced level of haemoglobin per cell

MIS-SENSE MUTATION the substitution of a base in DNA that leads to the incorporation of a different amino acid into a protein

MUTATION a heritable change in a gene

MUTATION RATE the frequency of mutations in each generation at a particular locus

NONSENSE MUTATION a mutation involving a base change in DNA and hence in messenger RNA that leads to premature termination of its protein product. A common cause of genetic disease

NUCLEATED RED CELL a red cell precursor that is found in the bone marrow but only in the peripheral blood in various forms of anaemia

NUCLEIC ACIDS deoxyribonucleic acid (DNA) and ribonucleic acid (RNA)

NUCLEOTIDE BASE one of the five nucleotides adenine, guanine, cytosine, thymine, and uracil that are the subunits of nucleic acid polymers

OLIGONUCLEOTIDE a short length of single-stranded polynucleotides, usually less than 30 nucleotides long

PEPTIDE a string of amino acids joined by peptide bonds

PEPTIDE BOND a bond formed between the amino group of one amino acid and the carboxyl group of another; hence, when a group of amino acids is joined together, one end of the resulting peptide chain will have a free amino group and the other a carboxyl group

PHENOTYPE the observable characteristics of an organism resulting from its genetic make-up and the environment

POLYMORPHISM the occurrence of two or more different alleles at a particular gene locus in a population, in cases where at least two alleles occur with frequencies of more than 1 per cent

RECESSIVE an allele that causes a phenotypic effect only when present in the homozygous state—that is, when it is inherited from both parents

RED BLOOD CELL the blood cells that carry haemoglobin and hence are involved in oxygenation of the tissues

RETICULOCYTE a red cell that has been recently released from the bone marrow and still contains residual nuclear material that can be demonstrated by a specific staining; increased numbers in the blood usually reflect a haemolytic anaemia

RIBONUCLEIC ACID (RNA) similar to DNA but with ribose as its sugar and uracil instead of thymine

RIBOSOME small particles made up of a particular form of RNA called ribosomal RNA that are involved in supporting the growing peptide chains during their synthesis on the messenger RNA template

SPLENOMEGALY enlargement of the spleen

STOP CODON the three bases of messenger RNA that encode the position for cessation of peptide chain synthesis

TRANSFER RNA (TRNA) a specific form of RNA that transports amino acids to the correct codon on messenger RNA; one end of this molecule is a triplet called the anti-codon that has three bases complementary to the messenger RNA codon for a particular amino acid; the other end holds the amino acid that is specified by the codon and hence the appropriate amino acid is transported to the correct position on the messenger RNA during peptide chain synthesis

BIBLIOGRAPHY AND FURTHER READING

This short bibliography is divided into two sections. The first provides the reader with further sources that cover the history of haematology and the thalassaemia and haemoglobin fields and also the related topics of the history of medical and molecular genetics. It also provides sources for further reading in the increasingly important field of the ethical and social issues of modern molecular genetics. Finally, those who wish to learn more about the work of individual scientists cited in the text are referred to the extensive review of the haemoglobin field edited by Steinberg and his colleagues. The second part of the bibliography consists of key or representative papers relating to the work described in each individual chapter.

Further Reading

History of Haematology

Wintrobe, M. M. (1980). *Blood, Pure and Eloquent*. New York: McGraw-Hill.

Wintrobe, M. M. (1985). *Hematology, the Blossoming of a Science: A Story of Inspiration and Effort*. Philadelphia: Lea and Febiger.

Zuelzer, W. W., and Nathan, D. G. (2009). Pediatric hematology in historical perspective. In S. H. Orkin et al., *Nathan and Oski's*

Hematology of Infancy and Childhood. 7th edn. Philadelphia: Saunders.

History of Thalassaemia and the Haemoglobin Field

Allison, A. C. (2002). The discovery of resistance to malaria by sickle cell heterozygotes. *Biochemistry, Molecular Biology and Education*, 30: 279–87.

Bannerman, R. M. (1961). *Thalassemia: A Survey of Some Aspects.* New York and London: Grune and Stratton.

Ferry, G. (2007). *Max Perutz and the Secret of Life.* London: Chatto and Windus.

Hager, T. (1995). *Force of Nature: The Life of Linus Pauling.* New York: Simon and Schuster.

Ingram, V. M. (2004). Sickle-cell anemia hemoglobin: the molecular biology of the first 'molecular disease'—the crucial importance of serendipity. *Genetics*, 167: 1–7.

Schechter, A. N. (2008). Hemoglobin research and the origins of molecular medicine. *Blood*, 112: 3927–38.

The Thalassemic Syndromes: A Symposium in Honour of Ezio Silvestroni and Ida Bianco (2001). Atti Dei Convegni Lincei, 164; Rome: Academia Nazionale Dei Lincei.

Weatherall, D. J. (1980). Toward an understanding of the molecular biology of some common inherited anemias: the story of thalassemia. In M. M. Wintrobe (ed.), *Blood, Pure and Eloquent*. New York: McGraw-Hill, 373–414.

Weatherall, D. J. (2001). Towards molecular medicine: reminiscences of the haemoglobin field, 1960–2000. *British Journal of Haematology*, 115: 729–38.

Weatherall, D. J. (2004). Thalassaemia: the long road from bedside to genome. *Nature Reviews Genetics*, 5: 625–31.

Weatherall, D. J., and Clegg, J. B. (2001). *The Thalassaemia Syndromes*. 4th edn. Oxford: Blackwell Science.

History of Medical and Molecular Genetics

Bearn, A. G. (1993). *Archibold Garrod and the Individuality of Man*. Oxford: Clarendon Press.

Harper, P. S. (2008). *A Short History of Medical Genetics*. Oxford: Oxford University Press.

McKusick, V. A. (1989). Forty years of medical genetics. *Jama*, 261: 3155–8.

McKusick, V. A. (2006). A 60-year tale of spots, maps, and genes. *Annual Reviews of Genomics and Human Genetics*, 7: 1–27.

Morange, M. (1998). *A History of Molecular Biology*. Cambridge, MA: Harvard University Press.

Ethical and Social Issues

Harris, J. (1998). *Clones, Genes and Immortality: Ethics and the Genetic Revolution*. Oxford: Oxford University Press.

Kevles, D. J. (1985). *In the Name of Eugenics*. Berkeley and Los Angeles: University of California Press.

Kevles, D. J., and Hood, L. (eds.) (1992). *The Code of Codes: Scientific and Social Issues in the Human Genome Project*. Cambridge, MA: Harvard University Press.

Walloo, K. (2001). *Dying in the City of the Blues: Sickle Cell Anemia and the Politics of Race and Health*. Chapel Hill, NC: University of North Carolina Press.

Extensive Review of the Haemoglobin Field and Bibliography Since 1960

Steinberg, M. H., Forget, B. G., Higgs, D. R., and Weatherall, D. J. (eds.) (2009). *Disorders of Hemoglobin*. 2nd edn. New York: Cambridge University Press.

Some Key References to Individual Chapters

Prologue

Partridge, G. (1998). *Alexandra Hospital*. Alexandra Hospital, Singapore Polytechnic Publication.

Weatherall, D. J., and Vella, F. (1960). Thalassaemia in a Gurkha family. *British Medical Journal*, 1: 1711–13.

Chapter 1

Bannerman, R. M. (1961). *Thalassemia: A Survey of Some Aspects*. New York and London: Grune and Stratton.

Bruce-Chwatt, L. J., and de Zulueta, J. (1980). *The Rise and Fall of Malaria in Europe: A Historico-Epidemiological Study*. Oxford: Oxford University Press.

Cappell, D. F. (1951). Myelophthisic Anaemia. In *Muir's Textbook of Pathology*. London: Arnold, 502.

Cardarelli, A. (1890). Nosografia della pseudo-leucemia splenica (infettiva) dei bambini; pelsocioord. *Rel. R. Acc. Med. Chir. Napoli*, Anno II: 17.

Cooley, T. B., and Lee, P. (1925). A series of cases of splenomegaly in children with anemia and peculiar bone changes. *Transactions of the American Pediatric Society*, 37: 29.

Cooley, T. B., Witwer, E. R., and Lee, P. (1927). Anemia in children with splenomegaly and peculiar changes in bones: report of cases. *American Journal of Diseases of Childhood*, 34: 347.

Crosby, W. H. (1980). The spleen. In M. M. Wintrobe (ed.), *Blood, Pure and Eloquent*. New York: McGraw Hill, 97–138.

Diggs, L. W. (1976). Dr. George Hoyt Whipple. *Johns Hopkins Medical Journal*, 139: 196–200.

Menke, W. T. (1973). Mediterranean anaemia in antiquity. *British Medical Journal*, 2: 489.

Osler, W. (1902). Anemia splenica. *Transactions of the Association of American Physicians*, 18: 429–61.

Rietti, F. (1925). Ittero emolitico primitivo. *Atti Acad. Sci. Med. Nar. Ferrara*, 2: 14.

Vaughan, J. (1936). *The Anaemias*. Oxford Medical Publications; Oxford: Oxford University Press.

Von Jaksch, R. (1889). Uber Leukaemia und leukocytose im kindesalter. *Wien Klin Wochenschr*, 2: 435.

Weatherall, D. J. (1980). Toward an understanding of the molecular biology of some common inherited anemias: the story of thalassemia. In M. M. Wintrobe (ed.), *Blood, Pure and Eloquent* (New York: McGraw-Hill), 373–414.

Weatherall, D. J., and Clegg, J. B. (2001). *The Thalassaemia Syndromes*. 4th edn. Oxford: Blackwell Science.

Whipple, G. H., and Bradford, W. L. (1932). Racial or familial anemia of children: associated with fundamental disturbances of bone and pigment metabolism (Cooley-Von Jaksch). *American Journal of Diseases of Childhood*, 44: 336–65.

Zaino, E. C. (1964). Paleontologic Thalassemia. *Annals of the New York Academy of Sciences*, 119: 402–12.

Chapter 2

Allison, A. C. (2002). The discovery of resistance to malaria by sickle cell heterozygotes. *Biochemistry, Molecular Biology and Education*, 30: 279–87.

Angelini, V. (1937). Primi risultati di ricerche ematologiche nei familiari di ammalati di anemia di Cooley. *Minerva Medica*, 28: 331–2.

Caminopetros, J. (1938). Recherches sur l'anémia érythroblastique des peuples de la Médterranée orientale. Premier Memoire: étude nosologique. *Annals of Medicine*, 43: 27.

Caminopetros, J. (1938). Recherches sur l'anémia érythroblastique infantile des peuples de la Mediterranée orientale. Étude anthropologique, étiologique et pathogénique. La Transmission héréditaire de la maladia. *Annals of Medicine*, 43: 104–25.

Chini, V., and Valeri, C. M. (1949). Mediterranean hemopathic syndromes. *Blood*, 4: 989–1013.

Cooley, T. B., and Lee, P. (1932). Erythroblastic anemia: additional comments. *American Journal of Diseases of Childhood*, 43: 705.

Garrod, A. E. (1909). Inborn Errors of Metabolism. In *The Croonian Lectures delivered before the Royal College of Physicians of London in June 1908*. London: Frowde, Hodder and Stoughton and the Royal College of Physicians of London.

Haldane, J. B. S. (1949). Disease and evolution. *Ricera Sci.*, 19: 2.

Haldane, J. B. S. (1949). The rate of mutation of human genes. *Proceedings of the VIII International Congress on Genetics and Hereditas*, 35: 267–73.

Lederberg, J. (1999). J. B. S. Haldane (1949) on infectious disease and evolution. *Genetics*, 153: 1–3.

Moncrieff, A., and Whitby, L. E. H. (1934). Anaemia of Cooley. *Lancet*, 2: 648.

Silvestroni, E., and Bianco, I. (1944–5). Microdrepanocitoanemia, in un sogetto di razza bianca. *Boll. Atti Acad. Med.*, 70: 347.

Silvestroni, E., and Bianco, I. (1945). Dimostrazione nell'uomo di una particolare anomalia ematologica costituzionale e rapporti fra questa anolalia e l'anemia microcitica costituzionale. *Policlinico*, 52: 1–44.

Silvestroni, E., Bianco, I., Montalenti, G., and Siniscalco, M. (1950). Frequency of microcythaemia in some Italian districts. *Nature*, 165: 682–3.

Valentine, W. N., and Neel, J. V. (1944). Hematologic and genetic study of transmission of thalassemia (Cooley's anemia: Mediterranean anemia). *Archives of Internal Medicine*, 74: 185–96.

Valentine, W. N., and Neel, J. V. (1948). A statistical study of the hematologic variables in subjects with thalassemia minor. *American Journal of Medical Science*, 215: 456–60.

Whipple, G. H., and Bradford, W. L. (1932). Racial or familial anemia of children: associated with fundamental disturbances of bone and pigment metabolism (Cooley-Von Jaksch). *American Journal of Diseases of Childhood*, 44: 336–65.

Wintrobe, M. M., Mathews, E., Pollack, R., and Dobyns, B. M. (1940). Familial hemopoietic disorder in Italian adolescents and adults resembling Mediterranean disease (thalassemia). *Journal of the American Medical Association*, 114: 1530–8.

Chapter 3

Ager, J. A. M., and Lehmann, H. (1958). Observations on some 'fast' haemoglobins: K, J, N and Bart's. *British Medical Journal*, 1: 929–31.

Bianco, I., Montalenti, G., Silvestroni, E., and Siniscalco, M. (1952). Further data on genetics of microcythemia or thalassaemia minor and Cooley's disease or thalassaemia major. *Annals of Eugenics*, 16: 299–314.

Chini, V., and Valeri, C. M. (1949). Mediterranean hemopathic syndromes. *Blood*, 4: 989–1013.

Conley, C. L. (1980). Sickle-cell anemia—the first molecular disease. In M. M. Wintrobe (ed.), *Blood, Pure and Eloquent*. New York: McGraw-Hill Book Company, 319–72.

Fessas, P., and Papaspyrou, A. (1957). A new fast haemoglobin associated with thalassemia. *Science*, 126: 1119.

Herrick, J. B. (1910). Peculiar elongated and sickle-shaped red blood corpuscles in a case of severe anemia. *Archives of Internal Medicine*, 6: 517–21.

Ingram, V. M. (1956). Specific chemical difference between the globins of normal human and sickle-cell anaemia haemoglobin. *Nature*, 178: 792–4.

Ingram, V. M., and Stretton, A. O. W. (1959). Genetic basis of the thalassemia diseases. *Nature*, 18: 1903–9.

Itano, H. A. (1953). Qualitative and quantitative control of adult hemoglobin. *American Journal of Human Genetics*, 5: 34–45.

Itano, H. A. (1957). The human hemoglobins: their properties and genetic control. *Advances in Protein Chemistry*, 12: 216–18.

Itano, H. A., and Pauling, L. (1961). Thalassaemia and the abnormal haemoglobins. *Nature*, 191: 398–9.

Itano, H. A., and Robinson, E. A. (1960). Genetic control of the α and β-chains of hemoglobin. *Proceedings of the National Academy of Sciences, USA*, 46: 1492.

Kunkel, H. G., Ceppellini, R., Müller-Eberhard, U., and Wolf, J. (1957). Observations on the minor basic hemoglobin component in blood of normal individuals and patients with thalassemia. *Journal of Clinical Investigation*, 36: 1615–25.

Kunkel, H. G., and Wallenius, G. (1955). New hemoglobin in normal adult blood. *Science*, 122: 288.

Marmont, A., and Bianchi, V. (1948). Mediterranean anaemia: clinical and haematological findings, and pathogenic studies in milder forms of disease (with report of cases). *Acta Haematologica*, 1: 428.

Minnich, V., Na-Nakorn, S., Chongchareonsuk, S., and Kochaseni, S. (1954). Mediterranean anemia: a study of 32 cases in Thailand. *Blood*, 9: 1–23.

Pauling, L. (1954). Abnormality of hemoglobin molecules in hereditary hemolytic anemias. In *The Harvey Lectures 1954–55*. New York: Academic Press.

Pauling, L., Itano, H. A., Singer, S. J., and Wells, I. G. (1949). Sickle-cell anemia, a molecular disease. *Science*, 110: 543–8.

Perutz, M. F., Rossman, M. G., Cullis, A. F., Muirhead, H., Will, G., and North, A. C. T. (1960). Structure of haemoglobin. *Nature*, 185: 416–22.

Rich, A. (1952). Studies on the hemoglobin of Cooley's anemia and Cooley's trait. *Proceedings of the National Academy of Sciences, USA*, 38: 187–96.

Rigas, D. A., Kohler, R. D., and Osgood, E. E. (1955). New hemoglobin possessing a higher electrophoretic mobility than normal adult hemoglobin. *Science*, 121: 372–5.

Silvestroni, E., and Bianco, I. (1944–5). Microdrepanocitoanemia, in un sogetto di razza bianca. *Boll. Atti Acad. Med.*, 70: 347.

Chapter 4

Atwater, J., Schwartz, I. R., Erslev, A. J., Montgomery, T. L., and Tocantins, L. M. (1960). Sickling of erythrocytes in a patient with thalassemia hemoglobin-I disease. *New England Journal of Medicine*, 263: 1215.

Baglioni, C. (1962). The fusion of two peptide chains in hemoglobin Lepore and its interpretation as a genetic deletion. *Proceedings of the National Academy of Sciences, USA*, 48: 1880–6.

Bank, A., and Marks, P. A. (1966). Excess α chain synthesis relative to β chain synthesis in thalassaemia major and minor. *Nature*, 212: 1198–1200.

Bank, A., and O'Donnell, J. V. (1969). Intracellular loss of free-α chains in β thalassaemia. *Nature*, 222: 295.

Brimhall, B., Hollan, S., Jones, R. T., Koler, R. D., Stocklen, Z., and Szelenyi, J. G. (1970). Multiple α chain loci for human hemoglobin. *Clinical Research*, 18: 184–6.

Clegg, J. B., Naughton, M. A., and Weatherall, D. J. (1965). An improved method for the characterization of human haemoglobin mutants: identification of $\alpha_2\beta_2$ 95Glu, haemoglobin N (Baltimore). *Nature*, 207: 945–7.

Clegg, J. B., Weatherall, D. J., Na-Nakorn, S., and Wasi, P. (1968). Haemoglobin synthesis in β-thalassaemia. *Nature*, 220: 664–8.

Conley, C. L., Weatherall, D. J., Richardson, S. N., Shepard, M. K., and Charache, S. (1963). Hereditary persistence of fetal hemoglobin: a study of 79 affected persons in 15 Negro families in Baltimore. *Blood*, 21: 261.

Dintzis, H. M. (1961). Assembly of the peptide chains of hemoglobin. *Proceedings of the National Academy of Sciences, USA*, 47: 247–50.

Edington, G. M., and Lehmann, H. (1955). Expression of the sickle-cell gene in Africa. *British Medical Journal*, 1: 1308–11.

Fessas, P. (1963). Inclusions of hemoglobin in erythroblasts and erythrocytes of thalassemia. *Blood*, 21: 21–32.

Gerald, P. S., Efron, M. L., and Diamond, L. K. (1961). A human mutation (the Lepore hemoglobinopathy) possibly involving two cistrons. *American Journal of Diseases of Childhood*, 102: 514.

Guidotti, G. (1962). In: Thalassaemia Conference on Hemoglobin, Arden House, Columbia University, New York.

Heywood, J. D., Karon, M., and Weissman, S. (1964). Amino acids: incorporation into alpha and beta-chain of hemoglobin by normal and thalassemic reticulocytes. *Science*, 146: 530–1.

Ingram, V. M. (1964). A molecular model for thalassemia. *Annals of the New York Academy of Sciences*, 119: 485–95.

Itano, H. A. (1965). The synthesis and structure of normal and abnormal haemoglobins. In J. H. P. Jonxis (ed.), *Abnormal Haemoglobins in Africa*. Oxford: Blackwell Scientific Publications, 3–16.

Jacob, F., and Monod, J. (1961). Genetic regulatory mechanisms in the synthesis of proteins. *Journal of Molecular Biology*, 3: 318.

Lie-Injo, L. E., and Jo, B. H. (1960). A fast moving haemoglobin in hydrops foetalis. *Nature*, 185: 698.

Marks, P. A., and Burka, E. R. (1964). Hemoglobin synthesis in human reticulocytes: a defect in globin formation in thalassemia major. *Annals of the New York Academy of Sciences*, 119: 513.

Marks, P. A., and Burka, E. R. (1964). Hemoglobins A and F: formation in thalassemia and other hemolytic anemias. *Science*, 144: 552–3.

Nance, W. E. (1963). Genetic control of hemoglobin synthesis. *Science*, 141: 123.

Nathan, D. G., and Gunn, R. B. (1966). Thalassemia: the consequences of unbalanced hemoglobin synthesis. *American Journal of Medicine*, 41: 815–30.

Pontremoli, S., Bargellesi, A., and Conconi, F. (1969). Globin chain synthesis in the Ferrara thalassemia population. *Annals of the New York Academy of Sciences*, 165: 253–69.

Smithies, O. (1959). An improved procedure for starch-gel electrophoresis: further variation in the serum proteins of normal individuals. *Biochemistry Journal*, 71: 585.

Sturgeon, P., and Finch, C. A. (1957). Erythrokinetics in Cooley's anemia. *Blood*, 12: 64–73.

Vaughan, J. (1948). *British Medical Journal*, 1: 35.

Wasi, P., Na-Nakorn, S., and Suingdumrong, A. (1964). Haemoglobin H disease in Thailand: a genetical study. *Nature*, 204: 907.

Weatherall, D. J. (1963). Abnormal haemoglobins in the neonatal period and their relationship to thalassaemia. *British Journal of Haematology*, 9: 265.

Weatherall, D. J., Clegg, J. B., and Naughton, M. A. (1965). Globin synthesis in thalassemia: an *in vitro* study. *Nature*, 208: 1061–5.

Chapter 5

Alter, B. P. (1990). Antenatal diagnosis. Summary of results. *Annals of the New York Academy of Sciences*, 612: 237.

Alter, B. P., Modell, C. B., Fairweather, D., Hobbins, J. C., Mahoney, M. J., Frigoletto, F. D., Sherman, A. N., and Nathan, D. G. (1976). Prenatal diagnosis of hemoglobinopathies: a review of 15 cases. *New England Journal of Medicine*, 295: 1437.

Angastiniotis, M. A., and Hadjiminas, M. G. (1981). Prevention of thalassaemia in Cyprus. *Lancet*, 1: 369–70.

Barry, M., Flynn, D. N., Letsky, E. A., and Risdon, R. A. (1974). Long-term chelation therapy in thalassaemia major: effect on liver iron concentration, liver histology and clinical progress. *British Medical Journal*, 1: 16–20.

Brittenham, G. M., Griffith, P. M., Nienhuis, A. W., McLaren, C. E., Young, N. S., Tucker, E. E., Allen, C. J., Farrell, D. E., and Harris, J. W. (1994). Efficacy of deferoxamine in preventing complications of iron overload in patients with thalassemia major. *New England Journal of Medicine*, 331: 567–73.

Canali, S., and Corbellini, G. (2002). Lessons from anti-thalassemia campaigns in Italy, before prenatal diagnosis. *Med. Secoli*, 14: 739–71.

Cao, A., Galanello, R., and Rosatelli, M. C. (1998). Prenatal diagnosis and screening of the haemoglobinopathies. *Clinical Haematology*, 11: 215–38.

Diamond, L. K. (1980). A history of blood transfusion. In M. M. Wintrobe (ed.), *Blood, Pure and Eloquent*. New York: McGraw Hill, 659–90.

Landsteiner, K. (1901). Uber agglutinationserscheinungen normalen menschlichen Blutes. *Wien Klin Wochenschr*, 14: 1132–4.

Loukopoulos, D., Hadji, A., Papadakis, M., Karababa, P., Sinopoulou, K., Boussiou, M., Kollia, P., Xenakis, M., Antsaklis, A., Mesoghitis, S., et al (1990). Prenatal diagnosis of thalassemia and of the sickle cell syndromes in Greece. *Annals of the New York Academy of Sciences*, 612: 226–36.

Modell, B., Petrou, M., Ward, R. H. T., Fairweather, D. V., Rodeck, C., Varnavides, L. A., and White, J. M. (1984). Effect of fetal diagnostic testing on birth-rate of thalassaemia major in Britain. *Lancet*, 2: 1383–6.

Modell, C. B., and Berdoukas, V. A. (1984). *The Clinical Approach to Thalassaemia*. New York: Grune and Stratton.

Modell, C. B., and Bulyjenkov, V. (1988). Distribution and control of some genetic disorders. *World Health Statistics Quarterly*, 41: 209–18.

Olivieri, N. F., Nathan, D. G., MacMillan, J. H., Wayne, A. S., Liu, P. P., McGee, A., Martin, M., Koren, G., and Cohen, A. R. (1994). Survival of medically treated patients with homozygous β thalassemia. *New England Journal of Medicine*, 331: 574–8.

Pippard, M. J., Callender, S. T., Letsky, E. A., and Weatherall, D. J. (1978). Prevention of iron loading in transfusion-dependent thalassaemia. *Lancet*, 1: 1178–80.

Propper, R. D., Cooper, B., Rufo, R. R., Nienhuis, A. W., Anderson, W. F., Bunn, H. F., Rosenthal, A., and Nathan, D. G. (1977). Continuous subcutaneous administration of deferoxamine in patients with iron overload. *New England Journal of Medicine*, 297: 418.

Sephton-Smith, R. (1964). Chelating agents in the diagnosis and treatment of iron overload. *Annals of the New York Academy of Sciences*, 119: 776.

Silvestroni, E., Bianco, I., Graziani, B., Carboni, C., and D'Acra, S. U. (1978). First premarital screening of thalassaemia carriers in intermediate schools in Latium. *Journal of Medical Genetics*, 15: 202–7.

Silvestroni, E., Bianco, I., Graziani, B., Carboni, C., Valente, M., Lerone, M., and D'Arca, S. U. (1981). Screening delle microcitemie nella popolazione scolastica del Lazio. *Minerva Medica*, 72: 677–83.

Stamatoyannopoulos, G. (1973). Problems of screening and counselling in the hemoglobinopathies. In A. G. Motulsky and W. Lenz (eds.), *IVth International Congress on Birth Defects*. Amsterdam and Vienna: Excerpta Medica, 268–76.

Wolman, I. J. (1964). Transfusion therapy in Cooley's anemia: growth and health as related to long-range hemoglobin levels, a progress report. *Annals of the New York Academy of Sciences*, 119: 736–47.

Wolman, I. J., and Ortolani, M. (1969). Some clinical features of Cooley's anemia patients as related to transfusion schedules. *Annals of the New York Academy of Sciences*, 165: 407.

Chapter 6

Antonarakis, S. E., Boehm, C. D., Giardina, P. V. J., and Kazazian, H. H. (1982). Non random association of polymorphic restriction sites in the β-globin gene complex. *Proceedings of the National Academy of Sciences, USA*, 79: 137–41.

Fessas, P. (1961). The beta-chain thalassaemias. In H. Lehmann and K. Betke (eds.), *Haemoglobin Colloquium, Vienna*. Stuttgart: Thieme, 90.

Hershko, C. (1989). Mechanisms of iron toxicity and its possible role in red cell membrane damage. *Seminars in Hematology*, 26: 277–85.

Hershko, C., Link, G., and Cabantchik, I. (1998). Pathophysiology of iron overload. *Annals of the New York Academy of Sciences*, 850: 191–201.

Jeffreys, A. J. (1979). DNA sequences in the $^{G}\gamma$-, $^{A}\gamma$-, δ- and β-globin genes of man. *Cell*, 18: 1.

Kan, Y. W., and Dozy, A M. (1978). Polymorphisms of DNA sequence adjacent to human β-globin structural gene: relation to sickle mutation. *Proceedings of the National Academy of Sciences, USA*, 75: 5631.

Kan, Y. W., Holland, J. P., Dozy, A. M., and Varmus, H. E. (1975). Demonstration of non-functional β-globin mRNA in homozygous β°-thalassemia. *Proceedings of the National Academy of Sciences, USA*, 72: 5140.

Kazazian, H. H., Jr., Orkin, S. H., Markham, A. F., Chapman, C. R., Youssoufian, H., and Waber, P. G. (1984). Quantification of the close association between DNA haplotypes and specific β-thalassaemia mutations in Mediterraneans. *Nature*, 310: 152–4.

Milner, P. F., Clegg, J. B., and Weatherall, D. J. (1971). Haemoglobin H disease due to a unique haemoglobin variant with an elongated α-chain. *Lancet*, 1: 729.

Olivieri, N. F., Rees, D. C., Ginder, G. D., Thein, S. L., Brittenham, G. M., Waye, J. S., and Weatherall, D. J. (1997). Treatment of thalassaemia major with phenylbutyrate and hydroxyurea. *Lancet*, 350: 491–2.

Orkin, S. H., Kazazian, H. H., Jr., Antonarakis, S. E., Goff, S. C., Boehm, C. D., Sexton, J. P., Waber, P. G., and Giardina, P. J. V. (1982). Linkage of β-thalassaemia mutations and β-globin gene polymorphisms with DNA polymorphisms in human β-globin gene cluster. *Nature*, 296: 627–31.

Orkin, S. H., Old, J. M., Weatherall, D. J., and Nathan, D. G. (1979). Partial deletion of β-globin gene DNA in certain patients with β°-thalassemia. *Proceedings of the National Academy of Sciences, USA*, 76: 2400–4.

Ottolenghi, S., Lanyon, W. G., Paul, J., Williamson, R., Weatherall, D. J., Clegg, J. B., Pritchard, J., Pootrakul, S., and Wong, H. B. (1974). The severe form of α thalassaemia is caused by a haemoglobin gene deletion. *Nature*, 251: 389–92.

Perrine, S. P., Ginder, G. D., Faller, D. V., Dover, G. H., Ikuta, T., Witkowska, E., Cai, S.-P., Vichinsky, E. P., and Olivieri, N. F. (1993). A short-term trial of butyrate to stimulate fetal-globin-gene expression in the β-globin disorders. *New England Journal of Medicine*, 328: 81–6.

Perrine, S. P., Greene, M. F., and Faller, D. V. (1985). Delay in the fetal globin switch in infants of diabetic mothers. *New England Journal of Medicine*, 312: 334–8.

Rachmilewitz, E. A. (1974). Denaturation of the normal and abnormal hemoglobin molecule. *Seminars in Hematology*, 11: 441–62.

Rachmilewitz, E. A., Shinar, E., Shalev, O., Galili, U., and Schrier, S. L. (1985). Erythrocyte membrane alterations in beta-thalassaemia. *Clinical Haematology*, 14: 163–82.

Rachmilewitz, E. A., and Thorell, B. (1972). Hemichromes in single inclusion bodies in red cells of beta thalassemia. *Blood*, 39: 794.

Schrier, S. L. (1994). Thalassemia: pathophysiology of red cell changes. *Annual Review of Medicine*, 45: 211–18.

Southern, E. M. (1975). Detection of specific sequences among DNA fragments separated by gel electrophoresis. *Journal of Molecular Biology*, 98: 503.

Spritz, R. A., Jagadeeswaran, P., Choudary, P. V., Biro, P. A., Elder, J. T., DeRiel, J. K., Manley, J. L., Gefter, M. L., Forget, B. G., and Weissman, S. M. (1981). Base substitution in an intervening sequence of a β^+ thalassemic human globin gene. *Proceedings of the National Academy of Sciences, USA*, 78: 2455–9.

Taylor, J. M., Dozy, A., Kan, Y. W., Varmus, H. E., Lie-Injo, L. E., Ganeson, J., and Todd, D. (1974). Genetic lesion in homozygous α-thalassaemia (hydrops foetalis). *Nature*, 251: 392–3.

Westaway, D., and Williamson, R. (1981). An intron nucleotide sequence variant in a cloned β⁺-thalassemia globin gene. *Nucleic Acids Research*, 9: 1777.

Wickramasinghe, S. N., Letsky, E., and Moffatt, B. (1973). Effect of α-chain precipitates on bone marrow function in homozygous β-thalassaemia. *British Journal of Haematology*, 25: 123–9.

Chapter 7

Alter, B. P. (1984). Advances in the prenatal diagnosis of hematologic diseases. *Blood*, 64, 329–40.

Appleyard, B. (1993). Our plunge in the gene pool. *Independent*.

Boehm, C. D., Antonarakis, S. E., Phillips, J. A., III, Stetten, G., and Kazazian, H. H., Jr. (1983). Prenatal diagnosis using DNA polymorphisms: report on 95 pregnancies at risk for sickle-cell disease or β-thalassemia. *New England Journal of Medicine*, 308: 1054–8.

Brittenham, G. M., Griffith, P. M., Nienhuis, A. W., McLaren, C. E., Young, N. S., Tucker, E. E., Allen, C. J., Farrell, D. E., and Harris, J. W. (1994). Efficacy of deferoxamine in preventing complications of iron overload in patients with thalassemia major. *New England Journal of Medicine*, 331: 567–73.

Charache, S., Barton, F. B., Moore, R. D., Terrin, M. L., Steinberg, M. H., Dover, G. J., Ballas, S. K., McMahon, R. P., and Castro, O. (1996). Fetal hemoglobin, hydroxyurea, and sickle cell anemia: clinical utility of a hemoglobin 'switching' agent. *Medicine*, 75: 300–6.

Charache, S., Dover, G., Smith, K., Talbot, C. C. J., Moyer, M., and Boyer, S. (1983). Treatment of sickle cell anemia with 5-azacytidine results in increased fetal hemoglobin production and is associated with nonrandom hypomethylation of DNA around the gamma-delta-beta-globin gene complex. *Proceedings of the National Academy of Sciences, USA*, 80: 4842–6.

Ghanei, M., Adibi, P., Movahedi, M., Khami, M. A., Ghasemi, R. L., Azarm, T., Zolfaghari, B., Jamshidi, H. R., and Sadri, R. (1997). Pre-marriage prevention of thalassaemia: report of a 100,000 case experience in Asfahan. *Public Health*, 111: 153–6.

Giardini, C. (1997). Treatment of β-thalassemia. *Current Opinion in Hematology*, 4: 79–87.

Hider, R. C., Kontoghiorghes, G. J., and Silver, J. (1982). UK Patent: GB-2118176.

Kan, Y. W., Golbus, M. S., and Dozy, A. M. (1976). Prenatal diagnosis of α-thalassemia: clinical application of molecular hybridization. *New England Journal of Medicine*, 295: 1165.

Kan, Y. W., Golbus, M. S., Klein, P., and Dozy, A. M. (1975). Successful application of prenatal diagnosis in a pregnancy at risk for homozygous beta thalassemia. *New England Journal of Medicine*, 292: 1096–9.

Lajtha, L. G. (1980). The common ancestral cell. In M. M. Wintrobe (ed.), *Blood, Pure and Eloquent*. New York: McGraw Hill, 81–96.

Lucarelli, G., Galimberti, M., Polchi, P., Angelucci, E., Baronciani, D., Giardini, C., Politi, P., Durazzi, S. M. T., Muretto, P., and Albertini, F. (1990). Bone marrow transplantation in patients with thalassemia. *New England Journal of Medicine*, 322: 417–21.

Lucarelli, G., and Weatherall, D. J. (1991). For debate: bone marrow transplantation for severe thalassaemia. *British Journal of Haematology*, 78: 300–3.

Old, J. M., Fitches, A., Heath, C., Thein, S. L., Weatherall, D. J., Warren, R., McKenzie, C., Rodeck, C. H., Modell, B., Petrou, M., and Ward, R. H. T. (1986). First-trimester fetal diagnosis for haemoglobinopathy: report on 200 cases. *Lancet*, 2: 763–7.

Old, J. M., Ward, R. H. T., Petrou, M., Karagozlu, F., Modell, B., and Weatherall, D. J. (1982). First-trimester fetal diagnosis for haemoglobinopathies: three cases. *Lancet*, 2: 1413–16.

Olivieri, N. F., Brittenham, G. M., Matsui, D., Berkovitch, M., Blendis, L. M., Cameron, R. G., McClelland, R. A., Liu, P. P., Templeton, D. M., and Koren, G. (1995). Iron-chelation therapy with oral deferiprone in patients with thalassemia major. *New England Journal of Medicine*, 332: 918–22.

Olivieri, N. F., Nathan, D. G., MacMillan, J. H., Wayne, A. S., Liu, P. P., McGee, A., Martin, M., Koren, G., and Cohen, A. R. (1994). Survival of medically treated patients with homozygous β thalassemia. *New England Journal of Medicine*, 331: 574–8.

Salihu, H. M. (1997). Genetic counselling among Muslims: questions remain unanswered. *Lancet*, 350: 1035–6.

Thomas, E. D., Buckner, C. D., Sanders, J. E., Papayannopoulou, T., Borgna-Pignatti, C., De Stefano, P., Sullivan, K. M., Clift, R. A., and Storb, R. (1982). Marrow transplantation for thalassaemia. *Lancet*, 2: 227–9.

Thompson, J., Baird, P., and Downie, J. (2001). *The Complete Text of the Report of the Independent Inquiry Commissioned by the Canadian Association of University Teachers*. Toronto: James Lorimer.

Wasi, P., and Fucharoen, S. (1995). The ethics of prenatal diagnosis in different religious and cultural contexts. In Y. Beyzard, B. Lubin, and J. Rosa (eds.), *Sickle Cell Disease and*

Thalassaemias: New Trends in Therapy. Colloque INSERM/John Libbey Eurotext Ltd., 234: 387–8.

Wintrobe, M. M. (1985). *Hematology, the Blossoming of a Science: A Story of Inspiration and Effort.* Philadelphia: Lea and Febiger.

Chapter 8

Allen, S. J., O'Donnell, A., Alexander, N. D. E., Alpers, M. P., Peto, T. E. A., Clegg, J. B., and Weatherall, D. J. (1997). α^+-thalassaemia protects children against disease due to malaria *and* other infections. *Proceedings of the National Academy of Sciences, USA*, 94: 14736–41.

Allison, A. C. (1954). Protection afforded by sickle-cell trait against subtertian malarial infection. *British Medical Journal*, 1: 290–4.

Allison, A. C. (2002). The discovery of resistance to malaria by sickle cell heterozygotes. *Biochemistry, Molecular Biology and Education*, 30: 279–87.

Angastiniotis, M., and Modell, B. (1998). Global epidemiology of hemoglobin disorders. *Annals of the New York Academy of Sciences*, 850: 251–69.

Bannerman, R. M. (1961). *Thalassemia: A Survey of Some Aspects.* New York and London: Grune and Stratton.

Brancati, C. (1961). Diffusione e frequencza della microcitemia e della anemie microcitemiche in Calabria. In *Proc. Conf. Il Problema Sociale della Microcitemia e del morbo di Coley.* Rome: It. Med. Soc., i. 64.

Carcassi, V., Ceppellini, R., and Pitzus, F. (1957). Frequenza della talassemia in quattro popolazioni sarde e suoi rapporti con la distribuzione dei gruppi sanguini e della malaria. *Boll. 1st. Sieroter. Mil.*, 36: 206.

Chatterjea, J. B. (1959). Haemoglobinopathy in India. In J. H. P. Jonxis and J. F. Delafresnaye (eds.), *Abnormal Haemoglobins*. Oxford: Blackwell Scientific Publications, 322–39.

Chernoff, A. I. (1959). The distribution of the thalassemia gene: a historical review. *Blood*, 14: 899–912.

Dronamraju, K. R. (2004). *Infectious Disease and Host-Pathogen Evolution*. New York: Cambridge University Press.

Flint, J., Hill, A. V. S., Bowden, D. K., Oppenheimer, S. J., Sill, P. R., Serjeantson, S. W., Bana-Koiri, J., Bhatia, K., Alpers, M. P., Boyce, A. J., Weatherall, D. J., and Clegg, J. B. (1986). High frequencies of α thalassaemia are the result of natural selection by malaria. *Nature*, 321: 744–9.

Friedman, M. J. (1978). Erythrocytic mechanism of sickle cell resistance to malaria. *Proceedings of the National Academy of Sciences, USA*, 75: 1994.

Jensen, J. B., and Trager, W. (1977). Plasmodium falciparum in culture: use of outdated erythrocytes and description of the candle jar method. *Journal of Parasitology*, 63: 883–6.

Kiple, K. F. (1993). *The Cambridge World History of Human Disease*. Cambridge: Cambridge University Press.

Livingstone, F. B. (1985). *Frequencies of Hemoglobin Variants*. New York and Oxford: Oxford University Press.

Modell, B., and Darlison, M. (2008). Global epidemiology of haemoglobin disorders and derived service indicators. *Bulletin of the World Health Organization*, 86: 480–7.

O'Shaughnessy, D. F., Hill, A. V. S., Bowden, D. K., Weatherall, D. J., Clegg, J. B., and with collaborators (1990). Globin genes in Micronesia: origins and affinities of Pacific Island peoples. *American Journal of Human Genetics*, 46: 144–55.

Pasvol, G., Weatherall, D. J., and Wilson, R. J. M. (1978). A mechanism for the protective effect of haemoglobin S against *P. falciparum*. *Nature*, 274: 701–3.

Silvestroni, E., and Bianco, I. (1944–5). Microdrepanocitoanemia, in un sogetto di razza bianca. *Boll. Atti Acad. Med.*, 70: 347.

Siniscalco, M., Bernini, L., Latte, B., and Motulsky, A. G. (1961). Favism and thalassaemia in Sardinia and their relationship to malaria. *Nature*, 190: 1179–80.

Todd, D. (1978). Genes, beans and Marco Polo. *University of Hong Kong Gazette*, 26: 1.

Vezzoso, B. (1946). Influenza della malaria sulla mortalita infantile per anemia con speciale riguardo al morbo di Cooley. *Riv. Malarial.*, 25: 61–77.

Weatherall, D. J., and Clegg, J. B. (2001). *The Thalassaemia Syndromes*. 4th edn. Oxford: Blackwell Science.

Weatherall, D. J., and Clegg, J. B. (2002). Genetic variability in response to infection: malaria and after. *Genes and Immunity*, 3: 331–7.

WHO (1994). *Guidelines for the control of haemoglobin disorders*. Geneva: World Health Organization.

Williams, T. N. (2006). Red blood cell defects and malaria. *Molecular Biochemistry and Parasitology*, 149: 121–7.

Chapter 9 Epilogue

Brown, M. S., and Goldstein, J. L. (1974). Familial hypercholesterolemia: defective binding of lipoproteins to cultured fibroblasts associated with impaired regulation of 3-hydroxy-3-methylglutaryl coenzyme A reductase activity. *Proceedings of the National Academy of Sciences USA*, 71: 788–92.

Collins, F. S., and McKusick, V. A. (2001). Implications of the Human Genome Project for Medical Science. *JAMA*, 285: 540–4.

Holtzman, N. A., and Marteau, T. M. (2000). Will genetics revolutionize medicine? *New England Journal of Medicine*, 343: 141–4.

Polani, P. E. (1990). *The Impact of Genetics on Medicine*. Harveian Oration to the Royal College of Physicians, London, 1988. London: Royal College of Physicians.

Vogel, F., and Motulsky, A. G. (1982). *Human Genetics: Problems and Approaches*. Berlin, Heidelberg, and New York: Springer-Verlag.

Weatherall, D. J. (1991). *The New Genetics and Clinical Practice*. 3rd edn. Oxford: Oxford University Press.

GENERAL INDEX